Just-in-Time

ALGEBRA AND TRIGONOMETRY FOR CALCULUS

Third Edition

Guntram Mueller

Ronald I. Brent

University of Massachusetts Lowell

PEARSON

Addison
Wesley

Boston San Francisco New York
London Toronto Sydney Tokyo Singapore Madrid
Mexico City Munich Paris Cape Town Hong Kong Montreal

Publisher	Greg Tobin
Senior Acquisitions Editor	William Hoffman
Senior Project Editor	Rachel S. Reeve
Editorial Assistant	Emily Portwood
Marketing Manager	Phyllis Hubbard
Marketing Coordinator	Heather Peck
Managing Editor	Karen Wernholm
Senior Production Supervisor	Jeffrey Holcomb
Copyeditor	Norma Emory
Proofreader	Kim Cofer
Indexer	Joe Wizda
Composition	Laura Wiegleb and Lynn L'Heureux
Art Illustration	George Nichols
Senior Manufacturing Buyer	Evelyn Beaton
Interior Design	Sandy Silva
Cover Design	Barbara T. Atkinson
Cover Image	Van Amburgh and Co.'s Menagerie, Circus and Colosseum Poster with Male Trapeze Artists © Swim Ink/CORBIS

Library of Congress Cataloging-in-Publication Data

Mueller, Guntram.
    Just-in-time algebra and trigonometry : for calculus / Guntram Mueller,
Ronald I. Brent.--3rd ed.
      p. cm.
    Includes bibliographical references and index.
    ISBN 0-321-26943-8
      1. Algebra. 2. Trigonometry. I. Brent, Ronald I. II. Title

QA154.3.M84 2004
512'.13--dc22

2004050759

Table of Contents

To the Student

Okay, you're in college, you're taking calculus, and as you flip through the pages of your calculus text, you see a lot of things that look suspiciously like algebra and trig. In fact, you see algebra on every page, along with some other odd characters like $\int$ and Δ and ∇. What's more, your professor lectures about the importance of algebra and trig in calculus, how they are the language of calculus, and you begin to pay a little more attention as you reflect on your own algebra and trig experience. You wonder: Is my background strong enough to handle calculus? Did they teach me the right things in high school? And did I learn them? Should I have spent more time on homework? (Ugh!) Will it all come back now to haunt me? (I need some air!) Where can I look up this stuff? And what do I need to look up anyway?

If this strikes a chord with you, take a close look at this book. It contains algebra and trig, and it is arranged in just the order that you'll need it, exactly at the moment when it comes up in your calculus course. That's the idea of *Just-in-Time Algebra and Trigonometry for Calculus.*

For example, take a look at the Table of Contents. When you take equations of circles and other conic sections in calculus, you need to know how to complete the square from algebra. There it is! When you take limits in calculus, you need to know how to factor from algebra. It's right there, just when you need it. No chasing around, no wading through piles of stuff you don't need. Just what you need, just when you need it, to allow you to calculate the limits. When you use implicit differentiation and need to solve some strange-looking equations of degree 1, here they are, to allow you to navigate with confidence and accuracy. Obviously, accuracy is important. Even profound ideas, without accuracy, can make bridges fall, buildings collapse, and rockets blow up. Not to mention how your boss might react!

What is the best way to use this book? Here's how. At any point in your calculus course:

1) Read the corresponding part of this book as directed by the Table of Contents, having pencil and paper at hand. If there are things you don't understand, or that just seem wrong (it happens!), put a question mark in the book and ask your professor in class. Make sure it all makes sense **to you**. **Your mind is the ultimate judge!**

2) When the presentation of the material and the examples make sense to you, do the odd-numbered exercises, **without using the book**, at least at first. It is important to become able to think for yourself in this context, as well as to get feedback on whether you can use the material correctly. Both of these goals are endangered when you do the exercises by looking for the example that "fits" the exercise and letting the example do the thinking for you. The answers to the odd-numbered exercises are in the back of the book.

Mathematics is not a subject like other subjects. It may not always be an easy subject to learn, and you should not be discouraged if it does not come right away, but with hard work, determination, and care, you **can** succeed in calculus. Millions have done it! Mathematics is unique unto itself in its representation of scientific thought. It is the language of science and technology. Mathematics is not mysterious, it can be learned, and for scientists and engineers it is absolutely essential.

For students using *Thomas' Calculus, Tenth Edition*, there is a *Just-In-Time* web site located at *http://www.mathxl.com/jito*. Students can take diagnostic tests online. The site contains practice and test exercises that are representative of the topics covered in *Just-In-Time Algebra and Trigonometry for Calculus*. For further information, contact your Addison-Wesley sales representative, or visit the Thomas' Calculus web site at *www.awl.com/thomas*.

To the Instructor

What calculus instructors have not, at one time or another, wrung their hands or pulled out their hair in despair over the level of preparedness of their students? This book is an attempt to try to salvage whatever is left of your hair.

As you can see from the Table of Contents, this book is organized by topic from most standard calculus courses. When limits are to be calculated, for example, the student needs factoring skills, so here are the factoring methods, all lined up at just the time the student needs them. There is too little room in calculus books for this material on factoring, and sending students to the library to look up factoring methods on their own does not produce the desired results.

Similarly, when studying conics, here's the method of completing the square. When studying the idea of the derivative, here is a review of rational operations on rational expressions. Doing implicit differentiation? Here's how to solve equations of degree 1, even strange-looking ones. The idea is that the student does not need to hunt through libraries and unfamiliar algebra texts to look for what he or she might need. Rather, it is all lined up here, "just-in-time," exactly when needed.

The material on exponentials, logs, and inverse trig functions is in the appendices, not because it is less important (far from it!), but because there is little uniformity on where it comes up in the different calculus books. So it's in the appendices, to be "taken as needed."

The back of the book contains the answers to the odd-numbered exercises only. The even-numbered answers are not given in the book, to provide some exercises that cannot be solved by "working toward the answer." The instructor may, however, find these online on the publisher's web site.

This book is intended to be used in one of two ways:

1) As a second text for calculus courses whose syllabi officially include a review of algebra and trigonometry. The organization of this book is a key feature: It can be used with almost all standard calculus texts, including the reform texts, in a manner that interleaves the topics from algebra and trigonometry with those from calculus. By timing certain topics in algebra and trigonometry to be given just before, or while, they are needed in calculus, we address the often-heard student complaint of irrelevance. For example, factoring is made more pertinent by tying it in with limit problems that can't be done except by factoring. This will enhance the motivation to learn factoring, a topic that can well use all the motivation it can get.

2)   As a companion to a standard calculus I course, for the majority of students, who in fact need a little well-timed help in algebra and trigonometry. This book is written in an easy style that can be understood by all students on their own, without input by the instructor. It is there as a reference, a guide, a handbook, a companion on a journey that is rewarding, but where the traveler could sometimes use a little support.

The idea for both uses is the same: the deft timing of the algebra and trigonometry review, topic-by-topic, just at the time that they are needed in calculus. It makes for a more relevant presentation of the review topic, which allows the student to be more interested in it because he or she is about to have to use it in calculus homework, that night or the next.

Acknowledgments

We wish to acknowledge all those who helped in this venture, first our wives Edie and Leor and our children Ariadne, Sarah, and Adam. Next, thanks to the many students who have helped point out weaknesses in earlier editions, and whose comments were vital to our revisions. Finally, we wish to thank our colleagues at the University of Massachusetts Lowell, as well as William Hoffman and the gang at Addison-Wesley.

Once in while you get shown the light, in the strangest of places, if you look at it right.

—*Robert Hunter*

Guntram Mueller

Ronald I. Brent

University of Massachusetts Lowell

Numbers and Their Disguises

Every number can be written in many different forms. For example, the numbers $\frac{8}{12}$, $\frac{10}{15}$, $\frac{3}{4 + \frac{\pi}{2\pi}}$, and even $\frac{|4x| + 4}{|6x| + 6}$ are all really just different ways of writing the number $\frac{2}{3}$. (Check it out; don't take **our** word for it.) For most purposes, the idea is to keep things as simple as possible, and just to use the form $\frac{2}{3}$.

Sometimes it is better to have a mathematical expression written as a product, other times it is better to have a sum; it all depends on what you need to do with it. In any event, **changing the form of an expression is something you have to do all the time**. Correctly! It's the nuts and bolts of mathematics, and is used in all the sciences and engineering, and even in economics and medicine.

1.1 Parentheses

Parentheses are ways of "packaging" or grouping numbers together.

EXAMPLE 1

$$5 - \left(1 + \frac{1}{2}\right) + (3 - 4) - \left(7 - \frac{1}{2}\right)$$

$$= 5 - \left(1\frac{1}{2}\right) + (-1) - \left(6\frac{1}{2}\right)$$

$$= 5 - \frac{3}{2} - 1 - \frac{13}{2}$$

$$= 5 - 1 - \frac{3}{2} - \frac{13}{2}$$

$$= 4 - \frac{16}{2} = 4 - 8 = -4 \quad \blacksquare$$

You may prefer to get rid of the parentheses. When you do that, don't forget: a) if there's a "+" immediately before the parentheses, leave all the signs of the terms inside as they are; b) if there is a "−" immediately before the parentheses, change all the signs of the terms inside. (These two rules are merely the results of the distributive law.)

EXAMPLE 2

$$5 - \left(1 + \frac{1}{2}\right) + (3 - 4) - \left(7 - \frac{1}{2}\right)$$

$$= 5 - 1 - \frac{1}{2} + 3 - 4 - 7 + \frac{1}{2}$$

$$= 5 - 1 + 3 - 4 - 7 - \frac{1}{2} + \frac{1}{2}$$

$$= -4 \quad \blacksquare$$

EXAMPLE 3   Simplify $3 - 2(8 + 1) - 3(5 - 7)$.

Solution Method 1: $3 - 2(9) - 3(-2) = 3 - 18 + 6 = -9$

Method 2: $3 - 16 - 2 - 15 + 21 = -9$ ■

Notice that in the second method in Example 3, the 8 and 1 were both multiplied by −2, and the 5 and −7 were multiplied by −3. This is based on the law of distributivity—namely, $a(b + c) = ab + ac$. Method 1 is easier in this case, but in many other cases Method 2 will be needed. **You've got to know both!**

In algebra, letters stand for numbers, so the laws of arithmetic apply to them in exactly the same way.

EXAMPLE 4 $xy - (2x - y) - 2y(1 - x)$

$$= xy - 2x + y - 2y + 2yx$$

$$= 3xy - 2x - y \quad \blacksquare$$

Notice that xy and $2yx$ add up to $3xy$.

EXAMPLE 5 $3x^2y - (x^2 - y^3) - 2y(x - y)$

$$= 3x^2y - x^2 + y^3 - 2xy + 2y^2 \quad \blacksquare$$

Notice the " + " in front of the y in Example 4 and the y^3 in Example 5, as well as the use of the distributive law in both examples. If the exponents in the last example are confusing you, skip ahead to Section 1.4 and then return here.

1.1 Exercises

Simplify:

1) $4 - (5 - 3) + 3(4 - 7) - 4(1 + 2)$ 2) $5 + (3 - 5) - (4 - 2) - (-5 - 3)$

3) $-2(3 - 4) + 4(5 + 3) - 3(2 - 6)$ 4) $2 - (3 - \frac{1}{4}) + (2.4 - 3.2) - (1.2 - 2.3)$

5) $4xy - (x - 2xy) - 2y(x - 1)$ 6) $(s - t) - (u - t) - (v - u) - (s - v)$

7) $2x(y - 3) - y(x + xy) + 2y(x + 1)$ 8) $x(y + z) - z(x + y) + 2y(x - z) - x(3y - 2z)$

9) $xy(x + y^2) - (2x^2y^2 - 2xy^3) - 2y^2(x^2 - y + xy)$ 10) $xy^2(x^2 + y^2) - 3x(2xy^2 - 2xy^3) + 2y^2(x - xy^2 - x^2)$

11) Consider simplifying the expression $-2(7 - 4) + 5(2 + 3) - 2(2 - 6)$ by a) first computing the expression within the parentheses and then by b) using the distributive law. Keep track of the total number of computations (additions, subtractions, or multiplications) in each method. Which way is "cheaper"?

1.2 Multiplying and Dividing Fractions
(We'll do adding and subtracting later.)

Multiplying fractions is the easiest manipulative task. The numerator is the product of the given numerators, and the denominator is the product of the given denominators. That is,

$$\frac{a}{b} \cdot \frac{c}{d} = \frac{a \cdot c}{b \cdot d}.$$

We use the symbol " $\cdot$ " to denote multiplication, although sometimes we will omit the symbol altogether.

EXAMPLE 1 a) $\dfrac{2}{3} \cdot \dfrac{5}{7} = \dfrac{2 \cdot 5}{3 \cdot 7} = \dfrac{10}{21}$

b) $\dfrac{1}{9} \cdot \left(-\dfrac{5}{8}\right) = \dfrac{1}{9} \cdot \dfrac{-5}{8} = \dfrac{1 \cdot (-5)}{9 \cdot 8} = \dfrac{-5}{72}$ ■

This rule can be extended to multiplying more than two fractions simply by multiplying across all the numerators and denominators.

EXAMPLE 2 a) $\dfrac{1}{4} \cdot \dfrac{7}{5} \cdot \dfrac{3}{8} = \dfrac{1 \cdot 7 \cdot 3}{4 \cdot 5 \cdot 8} = \dfrac{21}{160}$

b) $\dfrac{-1}{7} \cdot \dfrac{3}{8} \cdot \dfrac{-2}{\pi} = \dfrac{(-1)(3)(-2)}{7 \cdot 8 \cdot \pi} = \dfrac{6}{56\pi} = \dfrac{3}{28\pi}$ ■

Multiplying a number by 1 produces the same number, so, for example:
$\frac{5}{6} \cdot 1 = \frac{5}{6} \cdot \frac{4}{4} = \frac{20}{24}$. Going backward is usually more important: $\frac{20}{24} = \frac{5 \cdot 4}{6 \cdot 4} = \frac{5 \cdot \cancel{4}}{6 \cdot \cancel{4}} = \frac{5}{6}$. Here's the key point: you can cancel the factor 4 in the numerator and the denominator. More generally, if a number $c \neq 0$ is a factor of both the top and bottom of a fraction, it may be canceled. When all those common factors are canceled, the fraction is said to be in **lowest terms**.

EXAMPLE 3   Put $\frac{30}{84}$ into lowest terms.

Solution First factor the numerator and denominator as much as possible and then cancel all common factors.

$$\frac{30}{84} = \frac{\cancel{2} \cdot \cancel{3} \cdot 5}{2 \cdot \cancel{2} \cdot \cancel{3} \cdot 7} = \frac{5}{2 \cdot 7} = \frac{5}{14} \quad \blacksquare$$

WARNING: Make sure you cancel only those numbers that are factors of the entire top and the entire bottom. Don't be tempted to try "creative canceling." Consider the expression

$$\frac{3 + x^2}{3}.$$

Can you cancel the 3s ? **NO NO NO NO NO NO NO NO NO!!** Get the picture? The problem is that 3 is not a factor of the **entire** numerator, only of its first term. Hence you cannot simply cancel the 3s. But, if the expression had been, for example,

$$\frac{3 + 3x^2}{3},$$

then you could write:

$$\frac{3 + 3x^2}{3} = \frac{3(1 + x^2)}{3},$$

and now you could cancel the 3s to get $1 + x^2$.

Dividing by a fraction is done by inverting **that** fraction and multiplying:

$$\frac{\frac{a}{b}}{\frac{c}{d}} = \frac{a}{b} \cdot \frac{d}{c} = \frac{a \cdot d}{b \cdot c}.$$

For example,

$$\frac{\frac{-1}{3}}{\frac{5}{6}} = \frac{-1}{3} \cdot \frac{6}{5} = \frac{-6}{15} = \frac{-2}{5},$$

or you can cancel early:

$$\frac{-1}{{}_1\cancel{3}} \cdot \frac{\cancel{6}^2}{5} = \frac{-2}{5}.$$

EXAMPLE 4 Simplify the following expressions:

a) $\quad \dfrac{\frac{2}{3}}{\frac{3}{8}}$

b) $\quad \dfrac{\frac{5}{4}}{\frac{-10}{3}}$

Solution a) $\quad \dfrac{\frac{2}{3}}{\frac{3}{8}} = \dfrac{2}{3} \cdot \dfrac{8}{3} = \dfrac{16}{9}$

b) $\quad \dfrac{\frac{5}{4}}{\frac{-10}{3}} = \dfrac{\cancel{5}^1}{4} \cdot \dfrac{3}{-\cancel{10}_2} = -\dfrac{3}{8}$ ∎

EXAMPLE 5 Simplify the expression $\left(\dfrac{x+1}{y}\right) \cdot \left(\dfrac{-y+1}{x}\right)$.

Solution $\left(\dfrac{x+1}{y}\right) \cdot \left(\dfrac{-y+1}{x}\right) = \dfrac{(x+1) \cdot (-y+1)}{y \cdot x}$

$$= \dfrac{-xy + x - y + 1}{xy}$$

$$= \dfrac{1 - xy + x - y}{xy}$$ ∎

EXAMPLE 6 Simplify the expression $\dfrac{\frac{x^2 y}{z}}{\frac{xy^2}{z^3}}$.

Solution $\dfrac{\frac{x^2 y}{z}}{\frac{xy^2}{z^3}} = \dfrac{x^2 y}{z} \cdot \dfrac{z^3}{xy^2}$

$$= \dfrac{x^2 y z^3}{z x y^2} = \dfrac{x z^2}{y}$$ ∎

Again, if you are confused by the exponents, skip ahead to Section 1.4 and then return here.

In calculus, we use numbers called **real numbers**. These can be thought of as all decimal numbers, including those having an infinite number of digits. The set of real numbers corresponds to the set of points on the number line. Rational numbers are those real numbers that can be expressed as a quotient of two integers. (Integers are the numbers 0, 1, -1, 2, -2, etc.) For example, the numbers $\frac{1}{3}$, $\frac{4}{10}$, -5, $\frac{18}{7}$, $\sqrt{64}$, and 3.857 are all rational numbers. (Convince yourself of this by expressing them as the quotient of two integers.) All other real numbers are called irrational. The number $\sqrt{2}$ can be easily shown to be irrational. The number π is also irrational but that's much harder to prove.

1.2   Exercises

In Exercises 1–14, simplify and reduce to lowest terms. (Some answers may have more than one form.)

1) $\dfrac{5}{16} \cdot \dfrac{8}{10}$

2) $\dfrac{2\pi}{3} \cdot \dfrac{3\pi}{4}$

3) $\dfrac{-1}{3} \cdot \dfrac{-9}{5}$

4) $\dfrac{15}{28} \cdot \dfrac{12}{5} \cdot \dfrac{7}{8}$

5) $\dfrac{\frac{4}{75}}{\frac{8}{25}}$

6) $\dfrac{\frac{-7}{51}}{\frac{3}{12}}$

7) $\dfrac{\frac{3\pi}{7}}{\frac{2\pi}{3}}$

8) $\dfrac{x}{y} \cdot \dfrac{s}{u} \cdot \dfrac{y}{s} \cdot \dfrac{u}{x}$

9) $\dfrac{7x}{3y} \cdot \dfrac{3y + 2}{x}$

10) $\left(\dfrac{x + 2}{1 + y} \right) \cdot \left(\dfrac{1 - y}{x} \right)$

11) $\dfrac{\frac{xy}{w}}{\frac{xy - 2x}{w}}$

12) $\dfrac{xy}{wz} \cdot \dfrac{w^2 z}{x^2 y^2}$

13) $\dfrac{\frac{xy}{(x + y)}}{\frac{x^2 y}{(x + y)^3}}$

14) $\dfrac{\frac{xy}{(x - y)}}{\frac{x^2}{y} \cdot \frac{y^3}{x}}$

15)   Draw the number line, and locate the following numbers:

a) $\dfrac{-126}{106}$ b) $-3 + \dfrac{10}{27}$ c) $\dfrac{\pi}{2} - 1$

16)   Express the numbers -5, $\sqrt{64}$, and 3.857 as a quotient of two whole numbers.

1.3   Adding and Subtracting Fractions

Adding and subtracting fractions is easy when the denominators are all the same. For example,

$$\frac{2}{5} + \frac{17}{5} - \frac{4}{5} = \frac{2 + 17 - 4}{5} = \frac{15}{5} = 3.$$

But what if the fractions don't all have the same denominator? Then you first need to rewrite the fractions so that they all have the same denominator, called a **common denominator**. For example, if you want to add $\frac{2}{3} + \frac{4}{5}$, you can use a common denominator of 15. So

$$\frac{2}{3} + \frac{4}{5} = \frac{10}{15} + \frac{12}{15} = \frac{22}{15}.$$

Remember, to get a common denominator you can always use the product of the individual denominators ($3 \cdot 5 = 15$), but sometimes even a smaller number will do it, as long as both given denominators divide evenly into it. Then, to get the new numerator,

$$\frac{2}{3} = \frac{?}{15},$$

divide 3 into 15, to get 5, then multiply by 2, to get 10. Try it: $\frac{3}{7} = \frac{?}{42}$. (We're doing the opposite of canceling common factors.)

EXAMPLE 1   a)   Simplify $\dfrac{3}{5} + \dfrac{1}{2} - \dfrac{2}{3}$.

Solution $5 \cdot 2 \cdot 3 = 30$ will serve as a common denominator. So

$$\frac{3}{5} + \frac{1}{2} - \frac{2}{3} = \frac{18}{30} + \frac{15}{30} - \frac{20}{30}$$

$$= \frac{18 + 15 - 20}{30} = \frac{13}{30}$$

b) Simplify $\dfrac{1}{6} - \dfrac{1}{9}$.

Solution We could use $6 \cdot 9 = 54$ as a common denominator, but even 18 will do nicely because **both 6 and 9 divide evenly into 18**. So

$$\frac{1}{6} - \frac{1}{9} = \frac{3}{18} - \frac{2}{18} = \frac{1}{18}$$

c) Simplify $\dfrac{\frac{1}{2} + \frac{3}{4}}{\frac{1}{3} - \frac{1}{6}}$.

Solution $\dfrac{\frac{1}{2} + \frac{3}{4}}{\frac{1}{3} - \frac{1}{6}} = \dfrac{\frac{2+3}{4}}{\frac{2-1}{6}} = \dfrac{\frac{5}{4}}{\frac{1}{6}} = \dfrac{5}{4} \cdot \dfrac{6}{1} = \dfrac{15}{2}.$ ■

You will also have to be comfortable with adding and subtracting fractional expressions involving variables.

EXAMPLE 2 $\dfrac{xy}{z} + \dfrac{x}{5} = \dfrac{5xy}{5z} + \dfrac{xz}{5z} = \dfrac{5xy + xz}{5z}$ ■

EXAMPLE 3 $\dfrac{\frac{3a + 2b}{5ab}}{\frac{a}{b} - \frac{b}{a}} = \dfrac{\frac{3a + 2b}{5ab}}{\frac{a^2 - b^2}{ab}}$

$= \dfrac{3a + 2b}{5ab} \cdot \dfrac{ab}{a^2 - b^2}$

$= \dfrac{3a + 2b}{5a^2 - 5b^2}$ ■

1.3 Exercises

Express as a single fraction and simplify:

1) $\dfrac{1}{3} + \dfrac{1}{4}$

2) $\dfrac{7}{6} + \dfrac{5}{24}$

3) $\dfrac{2}{5} - \dfrac{1}{2} + \dfrac{1}{3}$

4) $\dfrac{1}{2} - \dfrac{1}{4} + \dfrac{1}{8} - \dfrac{1}{16}$

5) $\dfrac{1}{2} + \dfrac{4}{3} - \dfrac{2}{5} - \dfrac{3}{15}$

6) $\dfrac{5}{6} - \dfrac{4}{3} + \dfrac{2}{9} - \dfrac{3}{2}$

7) $\dfrac{\dfrac{1}{3} + \dfrac{2}{5}}{\dfrac{3}{2}}$

8) $\dfrac{\dfrac{1}{4} - \dfrac{2}{3}}{\dfrac{3}{2} - \dfrac{2}{5}}$

9) $\dfrac{\dfrac{2}{7} + \dfrac{1}{3}}{\dfrac{4}{3} + \dfrac{2}{5}} + \dfrac{1}{3}$

10) $\dfrac{2}{15}\left(\dfrac{1}{3} - \dfrac{2}{9}\right) - \dfrac{1}{3}\left(\dfrac{1}{6} - \dfrac{4}{9}\right) + \dfrac{1}{5}\left(\dfrac{2}{9} - \dfrac{1}{3}\right)$

11) $\dfrac{1}{x} + \dfrac{1}{y}$

12) $\dfrac{1}{y} - \dfrac{1}{x}$

13) $\dfrac{4}{x} - \dfrac{2}{y} + \dfrac{1}{z}$

14) $\dfrac{s}{tu} + \dfrac{t}{u} + \dfrac{u}{s}$

15) $\dfrac{1}{x} - \dfrac{x+1}{xy} + \dfrac{x-2}{xz}$

16) $\dfrac{\dfrac{1}{x} - \dfrac{1}{y}}{\dfrac{1}{x} + \dfrac{1}{y}}$

17) $\dfrac{\dfrac{1}{y} - \dfrac{x}{z}}{\dfrac{1}{z} - \dfrac{1}{x}}$

18) $\dfrac{1}{x}\left(\dfrac{1}{y} - \dfrac{1}{z}\right) - \dfrac{1}{y}\left(\dfrac{1}{x} - \dfrac{1}{z}\right)$

19) $\dfrac{\dfrac{1}{st} - \dfrac{1}{w}}{\dfrac{1}{tw} - \dfrac{2}{s}}$

20) $\dfrac{2}{x} + x\left(\dfrac{3}{xy} - \dfrac{4}{y}\right) - y\left(\dfrac{2}{x} - \dfrac{2}{xy}\right)$

21) $\dfrac{4yz}{x^2} - \dfrac{2z}{xy^2} + \dfrac{1}{xyz}$

22) $\dfrac{\dfrac{1}{x} - \dfrac{x}{y}}{\dfrac{2y}{x} + \dfrac{2x}{y}} + \dfrac{x-y}{xyz}$

1.4 Exponents

Positive whole number exponents are a simple mathematical notation to represent repeated multiplication by the same factor. But you can also have negative numbers and 0 as exponents. Consider this table.

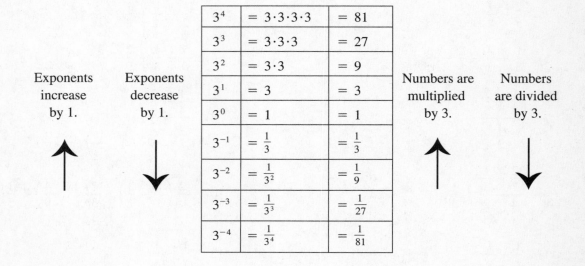

Exponents increase by 1.	Exponents decrease by 1.				Numbers are multiplied by 3.	Numbers are divided by 3.
		3^4	$= 3 \cdot 3 \cdot 3 \cdot 3$	$= 81$		
		3^3	$= 3 \cdot 3 \cdot 3$	$= 27$		
		3^2	$= 3 \cdot 3$	$= 9$		
		3^1	$= 3$	$= 3$		
↑	↓	3^0	$= 1$	$= 1$	↑	↓
		3^{-1}	$= \frac{1}{3}$	$= \frac{1}{3}$		
		3^{-2}	$= \frac{1}{3^2}$	$= \frac{1}{9}$		
		3^{-3}	$= \frac{1}{3^3}$	$= \frac{1}{27}$		
		3^{-4}	$= \frac{1}{3^4}$	$= \frac{1}{81}$		

See the pattern? Let's summarize: $3^n = \underbrace{3 \cdot 3 \cdot 3 \cdot \ldots \cdot 3 \cdot 3}_{n \text{ factors}}$, $3^{-n} = \dfrac{1}{3^n}$, and $3^0 = 1$.

This works not only for the number 3; for any number a and any positive integer n, we have:

$$a^n = \underbrace{a \cdot a \cdot a \cdot \ldots \cdot a \cdot a}_{n \text{ factors}}$$

$$a^{-n} = \frac{1}{a^n} \text{ for } a \neq 0,$$

and

$$a^0 = 1 \text{ for } a \neq 0.$$

In all of these rules, the number a is called the **base**, while the number n is called the **exponent**. There are several ways to simplify complicated expressions. Here are the laws of exponents that will help you. Know what they say, **and what they don't say**.

Laws of Exponents

Let m and n be any positive or negative integers, or 0. Let a and b be any real numbers. Then in each of the following, if the expressions on both sides exist, we have:

1) $a^m \cdot a^n = a^{m+n}$

2) $\dfrac{a^m}{a^n} = a^{m-n}$

3) $(a^m)^n = a^{m \cdot n}$

4) $(a \cdot b)^n = a^n \cdot b^n$

5) $\left(\dfrac{a}{b}\right)^n = \dfrac{a^n}{b^n}$

6) $\left(\dfrac{a}{b}\right)^{-n} = \left(\dfrac{b}{a}\right)^n$

(When might some of these expressions not exist?)

These rules are also true for values of m and n that are not integers. We will deal with such exponents in the next section.

EXAMPLE 1
a) $\dfrac{4^2 - 1}{3^3 - 2^2} = \dfrac{16 - 1}{27 - 4} = \dfrac{15}{23}$

b) $5^{-1} + 3^{-1} = \dfrac{1}{5} + \dfrac{1}{3} = \dfrac{3 + 5}{15} = \dfrac{8}{15}$

c) $\dfrac{3 \cdot 8^2}{9 \cdot 8^3} = \dfrac{3}{9} \cdot \dfrac{8^2}{8^3} = \dfrac{1}{3} \cdot \dfrac{8^2}{8^3} = \dfrac{1}{3} \cdot \dfrac{1}{8} = \dfrac{1}{24}$ ∎

EXAMPLE 2
$\dfrac{x^2 y^5}{x^{-3}} \div \dfrac{x^{-5} y^4}{x^3} = \left(\dfrac{x^2}{x^{-3}} y^5\right) \div \left(\dfrac{x^{-5}}{x^3} y^4\right)$

$= x^5 y^5 \div x^{-8} y^4$

$= \dfrac{x^5 y^5}{x^{-8} y^4} = x^{13} y$ ∎

EXAMPLE 3
$\dfrac{1}{a^3} - \left(\dfrac{1}{a^5} - \dfrac{1}{a^2}\right) = \dfrac{1}{a^3} - \dfrac{1}{a^5} + \dfrac{1}{a^2}$

$= \dfrac{a^2 - 1 + a^3}{a^5}$ ∎

EXAMPLE 4 $\left(\dfrac{x^{-2}}{x^8}\right)^{-2} = (x^{-2}x^{-8})^{-2}$

$$= (x^{-10})^{-2} = x^{20} \quad \blacksquare$$

EXAMPLE 5   a) $(x^2y^3)^5 = (x^2)^5(y^3)^5 = x^{10}y^{15}$

b) $\left(\dfrac{a^{-3}}{b}\right)^4 = \dfrac{(a^{-3})^4}{b^4} = \dfrac{a^{-12}}{b^4} = \dfrac{1}{a^{12}b^4}$

c) $\left(\dfrac{x^5}{y^3}\right)^{-2} = \left(\dfrac{y^3}{x^5}\right)^2 = \dfrac{(y^3)^2}{(x^5)^2} = \dfrac{y^6}{x^{10}} \quad \blacksquare$

EXAMPLE 6   Can we simplify $(x^2 + y^3)^7$?

Solution The laws of exponents are about quantities that are multiplied, divided, or exponentiated, but NOT ADDED OR SUBTRACTED. So there is no way of writing the above expression in a way that would be called simplified. (However, check out the Binomial Theorem in Appendix E for a different way of writing this expression.)   ■

EXAMPLE 7   a) $(x^2y^2)^{10} = x^{20}y^{20}$. Easy!

b) $(x^2 + y^2)^{10}$ can't be changed using exponent laws. See Example 6.

In any event, it is NOT equal to $x^{20} + y^{20}$.   ■

1.4 Exercises

In Exercises 1–9, simplify the expressions.

1) $\dfrac{4^{-1}5^2}{2^2 3^{-2}}$

2) $\dfrac{3^5 2^3}{4^2 3^3}$

3) $\dfrac{5^3}{3^{-1}5^2 + 4^{-1}5^3}$

4) $\dfrac{2^3}{4^{-1}2^2 + 3^{-1}2^3}$

5) $4^3 \left(\dfrac{1}{4}\right)^2 3^{-4}$

6) $\left(\dfrac{x^{-4}}{x^{-7}}\right)^{-2}$

7) $\dfrac{1}{2^{-3}} - \dfrac{1}{2} + \dfrac{1}{5^{-2}}$

8) $\dfrac{1}{3^{-2}} - \dfrac{1}{3} + \dfrac{1}{4^{-1}}$

9) $x^2y^{-2}z^3x^{-2}y^3z^5$

10)   Show by example that $(x^{-2} + y^{-2})^2 \neq x^{-4} + y^{-4}$; that is, find values for x and y so that the two sides are unequal for those values. (*Hint*: Just dive in and try some. Maybe you'll be lucky.)

Simplify using only positive exponents:

11) $\dfrac{x^{-1}y^2}{y^2x^{-2}}$

12) $\dfrac{(x^3y^{-2})^6}{(y^{-5}x^{-2})^{-3}}$

13) $\dfrac{(x^2y^{-3})^2}{(y^{-3}x^{-2})^{-2}}$

14) $\dfrac{x^4y^2}{x^{-3}} \div \dfrac{x^3y^{-2}}{y^5}$

15) $\dfrac{x^2y}{x^3} \div \dfrac{x^{-3}y^6}{y^4}$

16) $\dfrac{\dfrac{1}{x^2} - \dfrac{1}{y^3}}{\dfrac{1}{x^3} + \dfrac{1}{y^2}}$

17) $\dfrac{\dfrac{x^2y^{-3}}{3z^2} - \dfrac{z^{-3}y^{-3}}{3x^2}}{\dfrac{x^{-4}y^2}{3z^{-2}}}$

18) $\dfrac{2}{x^2} + x\left(\dfrac{3}{x^2y^2} - \dfrac{4x}{y}\right) - y\left(\dfrac{2y}{x^2} - \dfrac{2}{xy^2}\right)$

19) $(x^{-1} + y^{-1})^{-1}$

20) $\dfrac{\dfrac{x^3y^{-2}}{2z^5} - \dfrac{z^{-2}x^{-3}}{y^4}}{\dfrac{x^2y^{-2}}{2z^2}}$

21) $\left(\dfrac{x^{-2}}{x^{-3}}\right)^{-4}$

22)   The body-mass index, abbreviated BMI, is given by the formula $BMI = 703wh^{-2}$, where w is the person's weight in pounds and h is the person's height in inches. Find the BMI for Charles Barkley, who played in the 1992 Olympics at 6 ft 6 in. and 250 pounds.

1.5 Roots *(Also Called Radicals)*

DEFINITION

> We say that the number y is a **square root** of the number x if $y^2 = x$. A **cube root** of the number x is a number y such that $y^3 = x$, and so on. In general, if n is any positive integer, we say that the number y is an "nth root" of the number x if $y^n = x$. The number n is called the **order** of the root.

For Roots of Even Order

1)   The number 16 has two square roots, 4 and -4. The "radical" symbol $\sqrt{\ }$ means the **positive** square root always! So, $\sqrt{16} = 4$, but $\sqrt{16} \neq -4$. We say that 16 has two square roots: $\sqrt{16}$ and $-\sqrt{16}$ (i.e., 4 and -4). The situation is similar for higher even-order roots: Each positive number x has two even-order nth roots, denoted $\pm \sqrt[n]{x}$ where $\sqrt[n]{x}$ is always taken as the **positive** nth root.

For example: $\sqrt[4]{16} = 2$, since $2^4 = 16$; also $\sqrt[6]{64} = 2$, $\sqrt[4]{81} = 3$, and $-\sqrt{81} = -9$.

2)   Since no real number multiplied by itself an even number of times can produce a negative number, negative numbers have no real roots of even order. So $\sqrt{-4}$ is not a **real** number. (Do you know what it is?) Notice that we usually write $\sqrt{}$ instead of $\sqrt[2]{}$.

To sum up:

1)   Every positive number x has two real nth roots if n is even—namely, $\sqrt[n]{x}$ and $-\sqrt[n]{x}$.

2)   Only positive numbers, and 0, have even-order roots that are real numbers.

For Roots of Odd Order

Things are a little different with odd-order roots. All numbers have exactly one odd-order nth root, denoted $\sqrt[n]{x}$. The number 8 has exactly 1 real cube root, namely, 2. So, $\sqrt[3]{8} = 2$. But notice $\sqrt[3]{-8} = -2$, because $(-2)^3 = -8$, and $\sqrt[3]{-27} = -3$, because $(-3)^3 = -27$. Get it? Multiplying a negative number by itself an odd number of times produces a negative number!

   An alternative notation for $\sqrt[n]{a}$ is $a^{1/n}$. Both notations mean exactly the same thing. For example, $8^{1/3} = 2$, $25^{1/2} = 5$ (not -5), and $(-16)^{1/2}$ is not defined (as a real number).

DEFINITION

> We can also define **fractional exponents**: if $\frac{m}{n}$ is in lowest terms, and if n and a are such that $\sqrt[n]{a}$ makes sense, then we define $a^{m/n}$ to be $\left(\sqrt[n]{a}\right)^m$, which equals $\sqrt[n]{a^m}$. If a is much larger than 1, it is usually easier to take the root first; it keeps the numbers down.

EXAMPLE 1   a) $8^{2/3} = \left(\sqrt[3]{8}\right)^2 = 2^2 = 4$

b) $\left(\dfrac{-1}{27}\right)^{4/3} = \left(\sqrt[3]{\dfrac{-1}{27}}\right)^4 = \left(\dfrac{-1}{3}\right)^4 = \dfrac{1}{81}$

c) $(-32)^{4/5} = \left(\sqrt[5]{-32}\right)^4 = (-2)^4 = 16$ ∎

Laws of Exponents

Let r and s be any rational numbers. Let a and b be any real numbers. Then, in each of the following, if the expressions on both sides exist, they will be equal. (When might they not exist?)

1) $a^r \cdot a^s = a^{r+s}$

2) $\dfrac{a^r}{a^s} = a^{r-s}$

3) $(a^r)^s = a^{r \cdot s}$

4) $(ab)^r = a^r b^r$

5) $\left(\dfrac{a}{b}\right)^r = \dfrac{a^r}{b^r}$

6) $\left(\dfrac{a}{b}\right)^{-r} = \left(\dfrac{b}{a}\right)^r = \dfrac{b^r}{a^r}$

Remark (*Very Important*) Look at law #4. Let $r = \frac{1}{2}$. Then we have $(ab)^{1/2} = a^{1/2}b^{1/2}$, and so $\sqrt{ab} = \sqrt{a}\sqrt{b}$. In words, the square root of a product is the product of the square roots. From law #5 we have $\sqrt{\frac{a}{b}} = \frac{\sqrt{a}}{\sqrt{b}}$, or the square root of a quotient is the quotient of the square roots. Nowhere are there laws that say that $\sqrt{a + b}$ equals $\sqrt{a} + \sqrt{b}$, **because it doesn't**. The square root of a sum is not the sum of the square roots. Also $\sqrt{a - b}$ is not $\sqrt{a} - \sqrt{b}$.

We can generalize the above remark to nth-order roots in the following two root laws.

Root Laws

Let n be a positive integer. Let a and b be real numbers. Then, as long as both sides exist, we have:

1) $\sqrt[n]{ab} = \sqrt[n]{a}\sqrt[n]{b}$

2) $\sqrt[n]{\dfrac{a}{b}} = \dfrac{\sqrt[n]{a}}{\sqrt[n]{b}}.$

(That's all. There are no similar laws for addition or subtraction.)

1.5 Exercises

In Exercises 1–25, simplify the expressions as much as possible, using rational exponent notation where appropriate.

1) $\sqrt{144}$

2) $\sqrt{169}$

3) $\sqrt[3]{-64}$

4) $\sqrt{\dfrac{1}{9}}$

5) $\sqrt[5]{-32}$

6) $\sqrt[5]{-243}$

7) $\sqrt{\dfrac{4}{49}}$

8) $\sqrt[3]{\dfrac{8}{27}}$

9) $8^{5/3}$

10) $(-8)^{5/3}$

11) $(-32)^{2/5}$

12) $-(32)^{2/5}$

13) $\left(\dfrac{16}{9}\right)^{-3/2}$

14) $\left(\dfrac{25}{16}\right)^{-3/2}$

15) $(.01)^{-3/2}$

16) $(.008)^{4/3}$

17) $2^{5/3}2^{3/5}$

18) $8^{5/3}4^{3/2}$

19) $(3^{2/3})^{3/4}$

20) $\dfrac{2^{4/7}}{2^{3/2}}$

21) $3(27^{2/3})^{5/2}$

22) $\dfrac{(2^{1/3})^{2/5}}{\sqrt[5]{2}}$

23) $\sqrt{64x^8}$

24) $3^{1/3}9^{1/3} + 2^{1/3}16^{1/6}$

25) $\dfrac{x^{1/3}y^{2/5}}{\sqrt[3]{xy}}$

26) If $a > 0$ and $b > 0$, is $\sqrt{a^2b^2}$ the same as ab? Justify your answer.

27) If $a > 0$ and $b > 0$, is $\sqrt{a^2 + b^2}$ the same as $a + b$? Justify your answer.

28) If $a > 0$ and $b > 0$, is $\sqrt{a^2 - b^2}$ the same as $a - b$? Justify your answer.

29) Is $\sqrt[3]{x^3 - 8}$ the same as $x - 2$? Justify your answer.

30) If $x^2 + y^2 = 25$, can we conclude that $x + y = 5$? Why or why not?

Simplify if possible:

31) $\sqrt{25x^4}$

32) $\sqrt[3]{-27x^6}$

33) $(4x^6)^{3/2}$

34) $\left(\dfrac{9x^8}{16y^4}\right)^{-1/2}$

35) $(81x^2 - 4y^2)^{-1/2}$

1.6 Percent

Percent, represented by the symbol %, means "per hundred." So 5% of 400 means 5 one-hundredths of 400, which is $\left(\dfrac{5}{100}\right)(400) = 20$. In general, $x\%$ of y is $\left(\dfrac{x}{100}\right)(y)$.

EXAMPLE 1   Find 15% of 90.

Solution $\dfrac{15}{100} \cdot 90 = \dfrac{15 \cdot 9}{10} = \dfrac{135}{10} = 13.5$ ■

EXAMPLE 2   Find 1% of 320.

Solution $\dfrac{1}{100} \cdot 320 = \dfrac{320}{100} = 3.2$

Note that 1% of any number is always $\dfrac{1}{100} = .01$ of the number, so mov point over two places to the left! (For example, 1% of 4567 is 45.67 is 3.782.) ■

EXAMPLE 3   Find $\frac{1}{3}\%$ of 930.

Solution Since 1% of 930 is 9.3, $\frac{1}{3}\%$ of 930 would be

Alternatively: $\dfrac{\frac{1}{3}}{100} \cdot 930 = \dfrac{930}{300} = \dfrac{93}{30} = \dfrac{31}{10} =$

EXAMPLE 4   Find 250% of 150.

Solution $\dfrac{250}{100} \cdot 150 = \dfrac{250 \cdot 15}{10} = 25 \cdot 15 =$ ming

EXAMPLE 5   A CD usually sells for $15.99. Shower Records
How many CDs can you get for $20, and exactly
no sales tax)?

Solution A 60%-off sale means that the CDs
Now 40% of $15.99 is $\dfrac{40}{100} \cdot (\$15.99) = \dfrac{4 \cdot}{}$
up to the nearest cent). So you can buy three C

EXAMPLE 6   In 1995, the federal budget called for a 30% cut in the deficit to $147 billion. What would the deficit have been if it had not been cut?

Solution Let x = the uncut deficit.

So 70% of x is 147 billion, that is,

$$\frac{70}{100} \cdot x = 147 \text{ billion}.$$

Solving for x: $x = \frac{100}{70} \cdot (147 \text{ billion})$

$$= 210 \text{ billion}.$$

So the deficit would have been 210 billion dollars. ■

1.6 Exercises

1) Find 25% of 200.

Find 3.3% of 7.1.

The number 587 is 45% of what number?

The number 58 is 22% of what number?

he bookstore is having an inventory sale with a 25% reduction in prices. How much will a $16.99 sweatshirt cost ? How about a $1499.00 computer?

rdware store is having a sale with a 15% reduction in prices. How much will a $78 drill cost you?

40% cost reduction, a textbook that you purchased cost $20.40. What was the original price?

1/3-off cost reduction, your new printer cost $98. What was the original price?

re store is having a sale with a 20% reduction in prices. How much will a $239 table saw cost you?

cost reduction, your new skates cost $84. What was the original price?

the government was selling foreclosed land at 35% off. He went to look at a farm going for the of $97,500. Joe bought the farm. How much did he save?

$100,000 to start a new business. The loan had an interest rate of "prime plus two," with only the once a year. (Nice if you can get it.) The prime rate in the first year was 6%, and so her interest

rest did she pay at the end of the first year?

the second year rose to 9%. How much did she pay after the second year, assuming the prin- uched?

interest payment increase from the first year to the second?

tage increase?

1.7 Scientific Notation, Calculators, Rounding

Decimal notation is OK for many numbers, but for really large or small numbers it is a big pain. Take the 1995 deficit (please!): 210 billion = 210,000,000,000 (in decimal form). But notice that all the zeros really represent multiplication by 10 so that:

$$210{,}000{,}000{,}000 = 21 \times 10^{10} = 2.1 \times 10^{11}.$$

This last expression, using powers of 10 is called **scientific notation**. Notice that scientific notation calls for:

$$\pm \left[\begin{array}{l} \text{a number greater than or equal} \\ \text{to 1, but less than 10, written} \\ \text{in decimal form} \end{array} \right] \times 10^{\,\text{[some integer, positive or negative, or 0]}}.$$

For example, you would write .0000536 in scientific notation as 5.36×10^{-5}. You would write 6437.8 as 6.4378×10^3. Just count the number of positions you have to move the decimal point so that the resulting number is between 1 and 10. If you have to move the decimal point left n places, the exponent is n; if you move it n places to the right, the exponent is $-n$.

Remark   On most calculators, very large or small numbers are given in something like scientific notation. For example, the number 3.28×10^{-9} may look something like 3.2800000 E-09, or 3.28^{-09}.

EXAMPLE 1   The number of molecules in a mole (a mole??) of gas is called *Avogadro's number*. It is equal to 6.023×10^{23}. Write it in decimal (nonscientific) form.

Solution $6.023 \times 10^{23} = 602{,}300{,}000{,}000{,}000{,}000{,}000{,}000$

See why scientific notation is so handy! ∎

EXAMPLE 2   The speed of light (in a vacuum) is 186,000 miles per second. H
(the distance light travels in 1 year)? Express the answer in s
ed to three digits.

Solution First, the speed of light written in scient
there are 365 days a year, so there are

$$365 \text{ days} \times 24 \text{ (hr/day)} = 87$$

and

$$365 \text{ (days/yr)} \times 24 \text{ (hr/day)} \times 3600 \text{ (sec/h}$$

In scientific notation, this number is 3.1536×1

As you know, in each of these seconds, light travels 186,000 miles. So in 1 year, it travels $(31{,}536{,}000)(186{,}000)$ miles, and hence

$$1 \text{ light-year} = (3.1536 \times 10^7)(1.86 \times 10^5)$$

$$= (3.1536)(1.86) \times (10^7)(10^5)$$

$$= 5.865696 \times 10^{12}$$

$$\cong 5.87 \times 10^{12} \text{ miles.} \quad \blacksquare$$

Remark 1 This is in scientific notation since the first factor (5.87) is between 1 and 10. If it had not been between 1 and 10, you would have to write it differently. For example, if the product had ended up being 22.87×10^{12}, you would have to write it as 2.287×10^{13}.

Remark 2 Notice that when the number was **rounded** to three digits, the result was 5.87, not 5.86. That's because the next digit was ≥ 5, and so 5.87 is a closer approximation to the actual value than 5.86. Right?

Scientific and graphing calculators have a key called y^x. If you're calculating interest, that key is a lifesaver.

EXAMPLE 3 a) Calculate $(1.07)^{10}$.

b) If you invest $1000 at 7% interest, compounded annually, what is the value of your investment in 10 years? (Round your answer to the nearest penny.)

Solution a) Enter: 1.07

Press: y^x

Enter: 10

Press: =

The result on your calculator should be 1.967151357. It's probably unnecessary and a big nuisance to have all those digits. It may be fine to round it to 1.967 or 1.97, depending on what you are using it for.

b) Well, after 1 year the value would be

$$1000 + 7\% \text{ of } 1000 = 1000 + (0.07)(1000) = 1000(1.07).$$

So after 2 years, it would be

$$[1000(1.07)](1.07) = 1000(1.07)^2.$$

Similarly, after 10 years it would be

$$1000(1.07)^{10} = \$1967.15,$$

where the answer is rounded to the nearest penny. $\blacksquare$

1.7 | Exercises

1) Express the following numbers in scientific notation, rounded to three digits:

   a) 382935.9938   b) −0.000724   c) 3.000001   d) 200.001

2) Express the following numbers in scientific notation, rounded to four digits:

   a) 132435.9   b) −0.00219979

3) Compute the following and express in scientific notation, rounded to three digits. You may use your calculator.

   a) $(2.35 \times 10^5) \times (4.032 \times 10^2)$ b) $(-6.15 \times 10^{-20}) \times (5.032 \times 10^6)$

   c) $(-5.001 \times 10^{-2}) \times (-7.001 \times 10^{-99})$ d) $\dfrac{3.24 \times 10^2}{4.23 \times 10^3}$

   e) $\dfrac{-1.33 \times 10^{-2}}{7.9 \times 10^5}$ f) $(3.82 \times 10^{-1})^3$

4) Compute the following and express in scientific notation, rounded to three digits. You may use your calculator.

   a) $(1.35 \times 10^3) \times (4.452 \times 10^7)$ b) $\dfrac{5.24 \times 10^5}{1.02 \times 10^6}$

5) a) If you invest $2500, making 6% annually for 20 years, what will its value be?

   b) If the cost of living increased at a constant rate for these 20 years at 4% annually, then the "real" value of your investment, meaning your money's purchasing power, increases at 2% annually. What is the increase in purchasing power in 20 years? What is the percentage increase?

6) The earth orbits the sun in an orbit that is approximately a circle with a radius of 1.495979×10^8 km. What is the area of this circle? Express the result in scientific notation, rounded to five digits.

7) a) If the $24 that bought Manhattan Island on May 6, 1626, had been invested at 5%, what would it be worth on May 6, 1998? Express the answer in scientific notation, rounded to two digits.

   b) If the interest rate had been 6%, what would be the value of the investment in 1998?

   c) Roughly speaking, the 1998 value at the 6% rate was how many times as big as the 1998 value at the 5% rate?

   d) What is the percentage increase?

8) The mass of the sun is 1.989×10^{30} kg and the mass of the earth is 5.976×10^{24} kg. So the sun is how many times as massive as the earth? Express the result in scientific notation, rounded to five digits.

1.8 Intervals

An **interval** is just a connected piece of the number line. For example, the set of all real numbers between 0 and 1, including 0 and 1 is called a **closed interval** and is denoted [0, 1]. If you don't mean to include 0 and 1, you have an **open interval** and express it by using round parentheses, (0, 1). If 0 is to be included, but not 1, you write [0, 1). Each of these intervals can also be shown on the number line, or expressed as a pair of inequalities. In general, if $a < b$:

Interval Notation	Number Line	Inequalities
$[a, b]$		$a \le x \le b$
(a, b)		$a < x < b$
$[a, b)$		$a \le x < b$
$(a, b]$		$a < x \le b$

Note Some books use a different notation for an interval on the number line.

Instead of

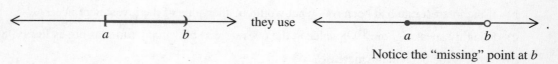

they use

Notice the "missing" point at b

All of these are called **finite intervals** because their length is finite. There are also **infinite intervals**. Here are some examples:

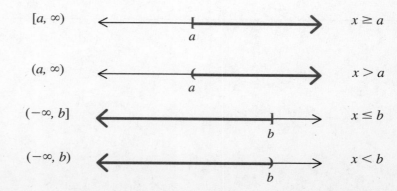

$[a, \infty)$		$x \ge a$
(a, ∞)		$x > a$
$(-\infty, b]$		$x \le b$
$(-\infty, b)$		$x < b$

DEFINITION │ Let A and B be two sets of objects of any sort.

a) The set of all objects that are in **both** A and B is called A **intersection** B, and is denoted $A \cap B$.

b) The set of all objects that are in **either** A **or** B **or both** is called A **union** B, and is denoted $A \cup B$.

c) The set that contains no elements is called the **empty set**, and is written as $\emptyset$.

These definitions can be applied to intervals, and they often are. Consider the following:

EXAMPLE 1

a) $(-\infty, 5) \cap (3, \infty)$ is the set of all numbers that are less than 5 and at the same time greater than 3. So it is the set $(3, 5)$.

b) Similarly, $(-\infty, 10] \cap (-1, \infty) = (-1, 10]$.

c) $(-\infty, 10) \cap (21, \infty) = \emptyset$, the empty set, since there are no numbers that are both less than 10 and at the same time greater than 21.

d) $(-\infty, 5) \cup (4, \infty) = (-\infty, \infty)$, the complete number line.

e) $(-\infty, 5) \cup (10, \infty)$ can't be written in a more compact form.

f) $(-\infty, 5) \cap [5, 10] = \emptyset$ ∎

In e), we see the use of union to describe sets of points on the number line that are not connected. It is often the case that we speak of a set of numbers consisting of two or more intervals, finite or infinite. This is when the union sign really comes in handy!

1.8 Exercises

1) Represent the following sets of numbers using interval and number line notation:

a) $-1 \leq x \leq 3$

b) $-1 < x \leq 3$

c) $-3 \leq x < 1$

d) $-3 \leq x \leq 4$

e) $-\dfrac{1}{2} < x \leq \sqrt{2}$

f) $\pi \leq x \leq 5$

g) $x < -4$

h) $0 < x$

i) $3 \leq x$

j) $3 - \pi \leq x$

k) $x < 5$

l) $x \leq 3$

2) Represent the following sets of numbers using interval and number line notation:

   a) $1 \le x < 15$ 　　　　　　　　　　　　b) $-2 \le x \le 2$

   c) $-5 < x \le -3$

3) Represent the following intervals using inequalities:

   a) $(3, 7)$ 　　　　　　　　　　　　　　　b) $(-4, -1]$

   c) $(-\infty, 19]$ 　　　　　　　　　　　　d) $[2, 10)$

   e) $[-2, -1]$

4) Represent the following intervals using inequalities:

   a) $(3, 7]$ 　　　　　　　　　　　　　　　b) $[-23, 15]$

   c) $(-\infty, 5]$

5) Simplify if possible:

   a) $(-\infty, 5) \cap [3, \infty)$ 　　　　　　　b) $(-\infty, 5) \cup [3, \infty)$

   c) $(-\infty, -2) \cap [-2, \infty)$ 　　　　　　d) $(-\infty, \infty) \cap [4, 7]$

   e) $[3, 5] \cap (10, \infty)$ 　　　　　　　　f) $(-\infty, 5] \cap [5, \infty)$

6) Sarah is trying to keep the water temperature in her chemistry experiment at 38°C. For the experiment to work, the relative error in the actual temperature must be less than 1%. Find the interval in which the actual temperature must lie, and write your solution using both inequalities and interval notation.

Completing the Square

2.1 Completing the Square

A quadratic expression in x is of the form $ax^2 + bx + c$. It's a polynomial of degree 2. One important way of changing the form of a quadratic is called **completing the square.** It is one of the most frequently used methods of changing the form of an expression to suit a particular purpose. It is used for graphing circles, ellipses, parabolas, and hyperbolas; for deriving the quadratic formula and integrating certain functions; and for many other purposes. You'll meet them soon enough. We'll do an example first, and then look at the general method.

EXAMPLE 1   Complete the square for $f(x) = x^2 + 8x + 12$.

Solution First, notice that the x^2 coefficient is 1. If it is not, it must be factored out. We take 8, the coefficient of x, then take half of the 8 to get 4, square the 4 to get 16, which we add (**and subtract**), to get:

$$f(x) = (x^2 + 8x + 16) + (12 - 16).$$

The first term is a perfect square. (That was the whole idea—and that's why it's called "completing the square.") So:

$$f(x) = (x + 4)^2 - 4. \quad \blacksquare$$

Here's the general method for completing the square of $f(x) = ax^2 + bx + c$:

a)   If $a \neq 1$, factor out the a from the first two terms to get

$$f(x) = a\left(x^2 + \frac{b}{a}x\right) + c.$$

b)   Take half of the coefficient of the resulting x-term, and square it.
c)   Add and subtract that number (inside the parentheses if $a \neq 1$).
d)   Rewrite $f(x)$ as the sum or difference of a perfect square and a number.

EXAMPLE 2   Complete the square for $f(x) = x^2 - 3x + 4$.

Solution Here, the coefficient of the x^2-term is 1, which means we can skip the first step. The x-coefficient is -3, half of it is $-\frac{3}{2}$ and squaring it gives $\frac{9}{4}$. Adding and subtracting $\frac{9}{4}$ gives us

$$f(x) = \left(x^2 - 3x + \frac{9}{4}\right) + \left(4 - \frac{9}{4}\right).$$

The first term is the perfect square of $x - \frac{3}{2}$, so:

$$f(x) = \left(x - \frac{3}{2}\right)^2 + \frac{7}{4}. \quad \blacksquare$$

EXAMPLE 3   Complete the square for $f(x) = 4x^2 + 20x - 100$.

Solution Now $a = 4$, so we must first factor it out of the first two terms:

$$f(x) = 4(x^2 + 5x) - 100.$$

The coefficient of the x-term is 5, halving and squaring it gives $\frac{25}{4}$ as the term to add and subtract, so

$$f(x) = 4\left(x^2 + 5x + \frac{25}{4} - \frac{25}{4}\right) - 100$$

(Notice the $\frac{25}{4}$ is subtracted **inside** the parentheses.)

$$= 4\left(x^2 + 5x + \frac{25}{4}\right) - 25 - 100$$

$$= 4\left(x + \frac{5}{2}\right)^2 - 125. \quad \blacksquare$$

Sometimes we need to complete the square in an equation. We may also need to complete the square in more than just one variable. Check the next example.

EXAMPLE 4   Complete the square in x and y for $x^2 - 4x + y^2 + 6y = 2$.

Solution Since half of the x-coefficient is -2, which when squared equals 4, we must first add 4 to both sides to complete the square in x. Next, half of the y-coefficient is 3, which when squared equals 9, so we add 9 to both sides giving

$$(x^2 - 4x + 4) + (y^2 + 6y + 9) = 2 + 4 + 9.$$

(Notice that instead of adding and subtracting on the left side, we added the same amount to both sides, which amounts to the same thing.)

So:

$$(x - 2)^2 + (y + 3)^2 = 15.$$

(By the way, this equation is that of a circle of radius $\sqrt{15}$ centered at the point $(2, -3)$.) ∎

The particular problem you solve will determine whether or not you choose to add the number to both sides of the equation, or just add and subtract the number on one side. The results are equivalent.

EXAMPLE 5   Complete the square in x and y for $4x^2 - 9y^2 + 8x + 18y - 25 = 0$.

Solution First regroup terms and factor out the coefficients of the quadratic terms:

$$4(x^2 + 2x) - 9(y^2 - 2y) - 25 = 0.$$

Now add and subtract appropriate constants: In this case both are 1.

$$4(x^2 + 2x + 1 - 1) - 9(y^2 - 2y + 1 - 1) - 25 = 0.$$

Upon simplifying:

$$4(x + 1)^2 - 4 - 9(y - 1)^2 + 9 - 25 = 0.$$

or

$$4(x + 1)^2 - 9(y - 1)^2 = 20. \quad ∎$$

EXAMPLE 6   Complete the square in x and y for $x^2 - \pi x + 2y^2 - y = 0$.

Solution For the quadratic in x, we need to add and subtract $\dfrac{\pi^2}{4}$, while for the y-terms we need to first factor out the 2 and then add and subtract $\dfrac{1}{16}$, giving

$$x^2 - \pi x + \frac{\pi^2}{4} - \frac{\pi^2}{4} + 2\left(y^2 - \frac{1}{2}y + \frac{1}{16} - \frac{1}{16}\right) = 0.$$

Simplifying gives

$$\left(x - \frac{\pi}{2}\right)^2 + 2\left(y - \frac{1}{4}\right)^2 = \frac{\pi^2}{4} + \frac{1}{8} = \frac{2\pi^2 + 1}{8}. \quad ∎$$

Completing the square is useful for analyzing and graphing the equations of circles, parabolas, and other conic sections. For example, the equation of the circle centered at (h, k), whose radius is r is

$$(x - h)^2 + (y - k)^2 = r^2,$$

according to your calculus book. So anytime you are given the equation of a circle, you can find its center and radius by transforming the equation into the form above, where you can read off the center (h, k), and the radius r.

EXAMPLE 7 Determine whether or not $x^2 + y^2 - 4x - 6y - 3 = 0$ is the equation of a circle. If so, find its center and radius.

Solution Let's complete the square in x and y. We rewrite as

$$(x^2 - 4x + 4) + (y^2 - 6y + 9) - 3 - 4 - 9 = 0$$
$$(x - 2)^2 + (y - 3)^2 - 16 = 0$$
$$(x - 2)^2 + (y - 3)^2 = 4^2$$

Hence, this is the equation of a circle of center $(2, 3)$ and radius 4. ■

EXAMPLE 8 Find the center and radius of the circle whose equation is

$$2x^2 + 2y^2 - 4x + 12y = -10.$$

Solution First divide by the coefficient of the x^2 and y^2 and then complete the square.

$$x^2 - 2x + y^2 + 6y = -5$$
$$(x^2 - 2x + 1) + (y^2 + 6y + 9) = -5 + 1 + 9$$
$$(x - 1)^2 + (y + 3)^2 = 5 = \left(\sqrt{5}\right)^2$$

So, the circle has center $(1, -3)$ and radius $\sqrt{5}$. ■

2.1 Exercises

1) Complete the square for the following expressions:

a) $f(x) = x^2 - 6x + 15$ b) $h(y) = y^2 + 5y$

c) $g(s) = s^2 + 2s - 8$ d) $k(x) = 2x^2 - 2x + 5$

e) $f(x) = 3x^2 - 7x + 1$ f) $w(x) = \pi x^2 + 2x$

2) Complete the square for the following expressions:

a) $f(x) = x^2 - 8x + 12$ b) $h(y) = y^2 + 14y$

c) $g(s) = s^2 + 3s - 6$ d) $k(x) = 4x^2 - 8x + 3$

3) Complete the square for the following equations:

   a) $x^2 - 3x - 17 = 0$

   b) $-3x^2 - 6x + 15 = 0$

4) Complete the square for the equation $x^2 - 5x - 1 = 0$.

5) Complete the square in both x and y for the following equations:

   a) $x^2 + 3x + 2y^2 - 8y = 0$

   b) $3x^2 + 6x - 2y^2 - 8y = -11$

   c) $-x^2 + 4x + y^2 - 16y = 40$

   d) $-9x^2 + 36x - 4y^2 - 8y = 0$

   e) $x^2 + y^2 - 6x + 10y + 34 = 0$

   The graph of this last example is called a **degenerate circle.** (Can you figure out why?)

6) Complete the square in both x and y for the following equations:

   a) $x^2 + 4x + 2y^2 - 6y = 0$

   b) $3x^2 - 12x - 2y^2 + 8y = -11$

   c) $-x^2 + 6x + y^2 - 8y = 40$

7) Find the center and radius of the circles represented by the following equations:

   a) $x^2 + y^2 - 4x - 2y = 11$

   b) $x^2 + y^2 - 6x + 4y - \pi^2 + 13 = 0$

   c) $2x^2 + 2y^2 + 4x + 8y - 20 = 0$

8) Find the center and radius of the circles represented by the following equations:

   a) $x^2 + y^2 - 6x - 8y = 0$

   b) $x^2 + y^2 - 10x + 12y + 12 = 0$

CHAPTER

Solving Equations

3.1 Equations of Degree 1 *(Linear Equations)*

You're asking: why are we doing this? This is the easiest thing in algebra. Well, it is, but it can get confusing when too many different variables are floating about. As long as you keep things straight, it really *is* easy. We'll start off simple, and each new example will bring up some new twist and how to deal with it.

The expression $ax + b$, with $a \neq 0$, is a polynomial of degree 1, and so the equation $ax + b = 0$ is called an **equation of degree 1**. Since the graph of the function $ax + b$ is a straight line (see Section 4.2), the equation $ax + b = 0$ is also called a **linear equation**.

EXAMPLE 1   Solve for x: $$4x - 16 = 0.$$

Solution First, get rid of the -16 by adding 16 to both sides:

$$4x - 16 + 16 = 0 + 16,$$

which gives $$4x = 16.$$

Now, get rid of the 4 in front of the x by dividing both sides by 4:

$$\frac{4x}{4} = \frac{16}{4},$$

giving $$x = 4.$$

Easy!

It's always a good idea to check your answer by substitution. That is, if we substitute $x = 4$ back into the original equation, it works! It is certainly true that

$$4 \cdot 4 - 16 = 0. \quad \blacksquare$$

EXAMPLE 2   Solve for x: $\dfrac{2}{3}x + 1 = 0$.

Solution Get rid of the 1 by subtracting it from both sides. This gives

$$\frac{2}{3}x + 1 - 1 = 0 - 1$$

or $$\frac{2}{3}x = -1.$$

Now, to isolate the variable x we need to divide by $\frac{2}{3}$. Instead, just like in Chapter 1, we multiply by its reciprocal, $\frac{3}{2}$:

$$\frac{3}{2} \cdot \frac{2}{3}x = \frac{3}{2} \cdot (-1).$$

This gives $$x = -\frac{3}{2}. \qquad \blacksquare$$

EXAMPLE 3   Solve for y: $\sqrt{2}y - 4 + \sqrt{2} = 0$.

Solution While this looks a little more complicated, y appears in only one place so you can start "peeling the onion." This is what we will call the method of stripping off terms, or factors, until the desired variable is exposed.

In this case, you first peel away the $-4 + \sqrt{2}$ by subtracting it from both sides to get

$$\sqrt{2}y = -\left(-4 + \sqrt{2}\right)$$

or $$\sqrt{2}y = 4 - \sqrt{2}.$$

The next layer to peel away is the $\sqrt{2}$ on the left side of the equation. Divide by $\sqrt{2}$ to get

$$y = \frac{4 - \sqrt{2}}{\sqrt{2}}.$$

We can simplify this answer as follows:

$$y = \frac{4 - \sqrt{2}}{\sqrt{2}} = \frac{4}{\sqrt{2}} - \frac{\sqrt{2}}{\sqrt{2}} = 2\sqrt{2} - 1. \qquad \blacksquare$$

In each of the preceding examples we obtained a solution by peeling the onion. The steps taken in stripping away the layers to expose the desired variable involved addition, subtraction, multiplication, or division of real numbers. Things can get a little more confusing when the equation involves more than the one variable you are solving for, but the idea is exactly the same. So, keep in mind what you have learned, and forge ahead!

EXAMPLE 4   Solve for x: $y^2x + w^2 = 0$.

Solution Now the equation has several variables. Remember the variable you wish to solve for and concentrate on isolating that one variable. Here, we wish to solve for x. Even though things seem a little more complicated, x still appears in only one place, so you can start peeling. In this case, first peel away the w^2 by subtracting it from both sides to get

$$y^2x = 0 - w^2$$

or $$y^2x = -w^2.$$

Next peel off the y^2 by division:

$$\frac{y^2x}{y^2} = \frac{-w^2}{y^2},$$

resulting in $$x = \frac{-w^2}{y^2}. \quad \blacksquare$$

Again, we solve the equation by stripping away the layers to expose the desired variable. Unlike the first three examples, the last example also had variables, w and y, in addition to the variable we were trying to solve for, x. These variables, however, just stand for real numbers, and so the mathematical steps taken to peel them away are exactly the same as those taken in the other examples.

Sometimes the variable you wish to solve for occurs more than once. In many cases this is not a problem because you can rewrite the equation in such a way that the variable occurs only once, and then solve as usual.

EXAMPLE 5   Solve for x: $4x + 3 = 2x + 1$.

Solution Here x occurs more than once, but if we subtract $2x$ from both sides, we get

$$2x + 3 = 1.$$

Notice that we have, in effect, moved all the terms containing x to the left side of the equation. Now, we subtract 3 from both sides to get

$$2x = -2,$$

where all the terms not containing x are now moved to the right side. Dividing by 2 gives the final answer:

$$x = \frac{-2}{2} = -1. \quad \blacksquare$$

EXAMPLE 6   Solve for x: $2x + 5y = 3x + y + 1$.

Solution Even though this appears more complicated because of the extra y-terms floating around, it really isn't any more difficult. Here, the x occurs more than once, so we subtract $3x$ from both sides, to get

$$-x + 5y = y + 1.$$

Notice that the terms containing x are all on the left side. Now, we subtract $5y$ from both sides:

$$-x = y + 1 - 5y,$$

which gives $-x = 1 - 4y.$

Notice that the terms not containing x are all on the right side. Dividing by -1 gives the final answer:

$$x = \frac{(1 - 4y)}{-1}$$
$$= \frac{1}{-1} - \frac{4y}{-1}$$
$$= -1 - (-4y)$$
$$= -1 + 4y$$
$$= 4y - 1. \quad \blacksquare$$

EXAMPLE 7   Solve for y: $\frac{2}{3}y + 2x - 1 = \frac{3}{4}y + x - \frac{1}{2}$.

Solution Here y occurs more than once, but we can proceed as in the last example. First we subtract $2x - 1$ from both sides:

$$\frac{2}{3}y = \frac{3}{4}y + x - \frac{1}{2} - (2x - 1),$$

which is the same as $\frac{2}{3}y = \frac{3}{4}y + x - \frac{1}{2} - 2x + 1,$

or $\frac{2}{3}y = \frac{3}{4}y + x - 2x + 1 - \frac{1}{2},$

and finally $\frac{2}{3}y = \frac{3}{4}y - x + \frac{1}{2}.$

Now, we can combine the two terms that contain y by subtracting $\frac{3}{4}y$ from both sides to get

$$\frac{2}{3}y - \frac{3}{4}y = \frac{3}{4}y - x + \frac{1}{2} - \frac{3}{4}y.$$

Canceling the $\frac{3}{4}y$'s on the right-hand side (RHS), and combining the y-terms on the left-hand side (LHS) gives:

$$\left(\frac{2}{3} - \frac{3}{4}\right)y = -x + \frac{1}{2}.$$

Remember how to subtract fractions: we get a common denominator, etc.

So

$$\frac{2}{3} - \frac{3}{4} = \frac{2 \cdot 4}{12} - \frac{3 \cdot 3}{12} = \frac{8}{12} = \frac{9}{12} = -\frac{1}{12}$$

and hence

$$-\frac{1}{12}y = -x + \frac{1}{2}.$$

Now, to expose the variable y, we divide by $-\frac{1}{12}$, that is, we multiply both sides by -12 to give the solution:

$$(-12) \cdot \left(-\frac{1}{12}\right)y = (-12) \cdot \left(-x + \frac{1}{2}\right)$$

or

$$y = (-12) \cdot (-x) + (-12) \cdot \left(\frac{1}{2}\right)$$

$$= 12x - 6. \quad \blacksquare$$

EXAMPLE 8   You bought two ties, three shirts (each of which cost twice as much as a tie), and a sweater that cost twice as much as a shirt, spending a total of $180. How much did the sweater cost?

Solution We let x = price of a tie, and express the other quantities in terms of x. So the price of a shirt is $2x$, and the price of a sweater is $4x$. Next, we express the fact that the total is $180 in the form of an equation, and solve it.

$$2(x) + 3(2x) + 4x = 180$$
$$2x + 6x + 4x = 180$$
$$12x = 180$$
$$x = 15$$

But, $15 is not the answer to the question! (Read the question.) The sweater cost $4x = 60$ (i.e., $60). $\blacksquare$

3.1 Exercises

1) Solve for x: $3x - 4 = 0$.

2) Solve for x: $2.34x - 20.23 = 0$.

3) Solve for x: $\dfrac{5}{12}x - 35 = 0$.

4) Solve for z: $\dfrac{9}{2}z + \dfrac{1}{3} = \dfrac{1}{6}$.

5) Solve for z: $\dfrac{4}{3}z - 1 = \dfrac{1}{10}$.

6) Solve for x: $1.23x - 3.01 = 0.67x + 2.56$.

7) Solve for x: $\dfrac{2}{3}x - 3 = \dfrac{1}{5}x + 2$.

8) Solve for y: $\dfrac{1}{4}y - \dfrac{1}{8} = \dfrac{1}{16} - 2y$.

9) Solve for y: $\dfrac{1}{2}y - \dfrac{1}{3} = \dfrac{1}{6} - 2y$.

10) The total cost for renting a copy machine for 60 months at $100 per month plus 5 cents a copy is $0.05x + 6000$ dollars, if x copies are made. If the same copy machine is purchased for $2000, then it costs 7 cents per copy for supplies and maintenance, for a total cost of $2000 + 0.07x$ dollars for x copies. Solve the equation $0.05x + 6000 = 2000 + 0.07x$ to find the number of copies for which the 5-year cost is the same as purchasing. Assuming the copier has no value after 5 years and assuming you make 2000 copies per month, is it cheaper to rent or purchase the machine?

11) Solve for x: $3y^2x + z^2 = 0$.

12) Solve for x: $2z^2x - z^3 = 1$.

13) Solve for y: $2zy + 3z + 1 = 0$.

14) Solve for y: $2zy + 4z + y = 2 - y$.

15) Solve for z: $2zy + 3z + 1 = 0$.

16) Solve for z: $2zy + 4z + y = 2 - y$.

17) Solve for x: $2y^2x - y^2 - (1 + 3y) = x$.

18) Solve for x: $x - y^2 + (2zy^2 + 2y)x = z^2x + z$.

19) You have 27 coins, which are made up of nickels, dimes, and quarters, whose total value is $3.30. You have three times as many quarters as dimes. How many nickels do you have?

20) A piece of wire 76-in. long is bent into the shape of a rectangle that is 2-in. longer than it is wide. What is the area of the rectangle?

21) One thousand dollars was invested, partly at 6% annual simple interest and partly at 4%. The total interest earned in the first year was $52. How much was invested at 6%?

22) Find the side of a square such that when each side is increased by 6 in., the area of the square is increased by 156 sq. in.

3.2   Equations of Degree 2 *(Quadratic Equations)*

The expression $ax^2 + bx + c$ with $a \neq 0$, is a polynomial of degree 2, and the equation $ax^2 + bx + c = 0$ is called an **equation of degree 2.** It is also called a **quadratic equation** because "quadratum" is Latin for "square." Notice that x occurs in two places, so you can't use the method of "peeling the onion." However, by using the method of completing the square, you can change the form of the equation in such a way that x occurs in only one place, and then peel away as usual. If you do that (see Appendix F), you will find that

$$x = \frac{-b \pm \sqrt{b^2 - 4ac}}{2a}.$$

This is the **quadratic formula.** It gives you the solution of that equation by simply substituting in the values of a, b, and c. No fuss, no bother, no completing the squares, no messy onion peels. It is one of the most useful formulas you will meet. MEMO-RIZE IT NOW!! If you don't, you will stray from the straight and narrow, and not even Elvis will be able to save you.

Remark   The "$\pm$" sign in the quadratic formula means that there are possibly two solutions, one if we pick the "$+$" sign and the other if we pick the "$-$" sign. Sometimes the two solutions are, in fact, equal. (When does this happen?)

EXAMPLE 1   Solve for x: $x^2 - 3x + 2 = 0$.

Solution Here $a = 1$, $b = -3$, and $c = 2$. So, according to the quadratic formula

$$x = \frac{-(-3) \pm \sqrt{(-3)^2 - 4 \cdot 1 \cdot 2}}{2 \cdot 1},$$

which simplifies to

$$x = \frac{3 \pm \sqrt{9 - 8}}{2} = \frac{3 \pm 1}{2}.$$

When the "$+$" sign is used, we get the solution

$$x = \frac{3 + 1}{2} = \frac{4}{2} = 2.$$

When the "$-$" sign is used, we get the solution

$$x = \frac{3 - 1}{2} = \frac{2}{2} = 1.$$

So there are two solutions: $x = 1$ and $x = 2$. Can you see how useful this formula is?   ∎

EXAMPLE 2   Solve for x: $2x^2 - 3x + 1 = 0$.

Solution Here $a = 2$, $b = -3$, $c = 1$, and $b^2 - 4ac = 1$. So, according to the quadratic formula

$$x = \frac{-(-3) \pm \sqrt{(-3)^2 - 4 \cdot 2 \cdot 1}}{2 \cdot 2},$$

which simplifies to

$$x = \frac{3 \pm \sqrt{9 - 8}}{4} = \frac{3 \pm 1}{4}.$$

The "+" sign gives $x = \dfrac{3 + 1}{4} = \dfrac{4}{4} = 1$.

The "−" sign gives $x = \dfrac{3 - 1}{4} = \dfrac{2}{4} = \dfrac{1}{2}$.

So there are two solutions: $x = 1$ and $x = \dfrac{1}{2}$. ∎

EXAMPLE 3   Solve for s: $s^2 + 4s + 4 = 0$.

Solution Don't let the s fool you, this is just a quadratic equation in s, and s takes the place of x in the preceding discussion. Here $a = 1$, $b = 4$, and $c = 4$. So,

$$s = \frac{-(4) \pm \sqrt{(4)^2 - 4 \cdot 1 \cdot 4}}{2 \cdot 1}$$

$$= \frac{-4 \pm \sqrt{16 - 16}}{2} = \frac{-4 \pm 0}{2} = -2.$$

In this case there is only one root. When this happens, the root is called **repeated**, or a **root of order 2**. (When speaking of equations, **root** is another word for **solution**.) It means that the original quadratic could be factored as $s^2 + 4s + 4 = (s + 2)^2$, and yes, you guessed it: Repeated roots occur when the quadratic is a perfect square. ∎

Sometimes you can solve a quadratic equation by "peeling the onion" more quickly than using the quadratic formula. This happens when the coefficient $b = 0$. Consider the next example.

EXAMPLE 4   Solve for x: $x^2 - 9 = 0$.

Solution Here $a = 1$, $b = 0$, and $c = -9$. Instead of using the quadratic formula we will "peel the onion." First we add 9 to both sides of the equation to get

$$x^2 = 9.$$

Now we take square roots of both sides of the equation, remembering that all positive numbers have two square roots. So

$$\pm x = \pm 3,$$

which means $x = 3$ or $x = -3$. ∎

Sometimes a quadratic equation can't be solved in terms of real numbers. Consider the following.

EXAMPLE 5 Solve for y: $y^2 + 2y + 2 = 0$.

Solution Here $a = 1$, $b = 2$, and $c = 2$, so that

$$y = \frac{-2 \pm \sqrt{4 - 8}}{2} = \frac{-2 \pm \sqrt{-4}}{2} = \frac{-2 \pm 2\sqrt{-1}}{2} = -1 \pm \sqrt{-1}.$$

The term $\sqrt{-1}$ is not defined as a real number because no real number when squared gives a negative number. So there are no real solutions. ■

If $b^2 - 4ac < 0$, as in the last example, there are no real solutions, meaning that there are no solutions that are real numbers. But wishful thinking is a great motivator, and in this spirit, mathematicians several centuries ago agreed to consider $\sqrt{-1}$ an "imaginary number." We represent $\sqrt{-1}$ by the symbol i (for imaginary). In this case the roots of the last equation could be written as $-1 \pm i$. Any number of the form $a + ib$ is called a **complex number**, and all four basic operations (addition, subtraction, multiplication, and division) carry over from real numbers, provided we remember $i^2 = -1$. (In the 1800s, complex numbers were removed from the realm of hocus-pocus and put onto a logically sound foundation—which means that today we can all sleep much easier.) Now check the numbers $-1 \pm i$ to see whether they are in fact solutions to the problem in Example 5.

EXAMPLE 6 Verify that $y = -1 \pm i$ are solutions to $y^2 + 2y + 2 = 0$.

Solution We will check each value, one at a time.

a) If $y = -1 + i$, then

$$y^2 = (-1 + i)^2 = (-1)^2 - 2i + i^2$$
$$= 1 - 2i - 1$$
$$= -2i$$
$$2y = 2(-1 + i) = -2 + 2i.$$

Hence $y^2 + 2y + 2 = -2i - 2 + 2i + 2 = 0$. So it works!

b) If $y = -1 - i$, then

$$y^2 = (-1 - i)^2 = (-1)^2 + 2i + (-i)^2$$
$$= 1 + 2i - 1$$
$$= +2i$$
$$2y = 2(-1 - i) = -2 - 2i.$$

Hence $y^2 + 2y + 2 = 2i - 2 - 2i + 2 = 0$. ■

Remark These examples demonstrate the three possible outcomes when solving quadratic equations. The number $b^2 - 4ac$ is called the **discriminant**, since it allows you to "discriminate" among the following three cases that indicate what types of roots you will have:

a) $b^2 - 4ac > 0$: In this case, the quadratic equation has two real and distinct roots—that is, two roots that are different numbers. This was the situation in Example 1.

b) $b^2 - 4ac = 0$: Here, the square root in the solution disappears, leaving one real repeated root. This happened in Example 3.

c) $b^2 - 4ac < 0$: In this last case, the roots are complex numbers, as was the case in Example 5.

It is possible that in your travels you will meet up with a quadratic equation in x where the coefficients a, b, and c are functions of some other variables. Take the following example.

EXAMPLE 7 Solve for x: $zx + 2yx^2 + yx + z^2 - y^2 = 0.$

Solution On first appearance this seems formidable, but by writing the x^2-term first, then the x-term, and then the term without x, we get

$$(2y)x^2 + (z + y)x + (z^2 - y^2) = 0.$$

We can see that this is a quadratic equation in x with $a = 2y$, $b = z + y$, and $c = z^2 - y^2$. Hence substituting these values into the quadratic formula gives

$$x = \frac{-(z + y) \pm \sqrt{(z + y)^2 - 8y(z^2 - y^2)}}{4y}. \quad \blacksquare$$

3.2 Exercises

Find the solutions, both real and complex, for Exercises 1–18.

1) $x^2 + 5x + 4 = 0$

2) $g^2 - 2g - 8 = 0$

3) $x^2 + 6x + 9 = 0$

4) $2y^2 + y - 1 = 0$

5) $y^2 - 8y + 16 = 0$

6) $\dfrac{x^2}{4} + 2x + 1 = 0$

7) $y^2 - 2y + 2 = 0$

8) $3y^2 - 4 = 0$

9) $x^2 + 3x = 4$

10) $x^2 + 7x + 4 = 0$

11) $g^2 - 4g - 5 = 0$

12) $8x^2 + 6x + 1 = 0$

13) $2y^2 + 7y + 4 = 0$

14) $y^2 - \dfrac{2y}{15} - \dfrac{1}{15} = 0$

15) $\dfrac{x^2}{3} + 2x - 1 = 0$

16) $y^2 + 3 = 0$

17) $x^2 + 2x = 4$

18) $3y^2 - 1.8y - 1.2 = 0$

19) Solve $\dfrac{1}{2}x^2 + yx - y^2 = 0$ for x.

20) Solve $\dfrac{x^2}{2} - 2y^2x + y^4 = 0$ for x.

21) Solve $\dfrac{1}{4}x^2 - (y + z)x + z^2 + y^2 = 0$ for x.

22) Solve $\dfrac{1}{4}x^2 - (y + z)x + z^2 + y^2 = 0$ for z.

23) A football field is 100-yd long from goal line to goal line and 160-ft wide. If a player ran diagonally across the field from one goal line to the other goal line, how far did he or she run?

24) The width of a rectangular gate is 2 ft more than its height. If a diagonal board of length 8 ft is used for bracing, then what are the dimensions of the gate?

25) The path of a projectile is given by the equation $y = 2x - 0.03x^2$, where x is the horizontal distance and y is the vertical distance from the initial point, in feet. a) Find x when $y = 32$ ft. b) Physically speaking, what does it mean to have two solutions?

26) The volume of a cube is increased by 61 cu in. by increasing each edge of the cube by 1 in. Find the edge of the original cube.

27) A rectangle 6″ × 10″ is to be reduced in size by cutting two strips of equal width, one from the length, the other from the width, so that the area is reduced by 15 sq in. How wide should the strips be?

3.3 Solving Other Types of Equations

Some of the methods that we have used to solve equations of degree 1 and 2 can also be used to solve other kinds of equations. There are four basic tools in our equation solving toolbox:

1) *Peeling the Onion,* for any equation where the variable that we wish to solve for occurs in only one place, or any equation that can be put into that form.

2) *The Quadratic Formula,* for equations that are of "quadratic type," meaning that a well-chosen substitution can change the given equation into an ordinary quadratic equation.

3) *The Zero-Factor Property (ZFP),* which allows us to solve equations of the form $f(x) = 0$, by factoring $f(x)$, and setting each factor equal to zero.

4)   *Graphing Calculator Techniques,* which can be used to solve the equation $f(x) = 0$, by graphing $f(x)$, and seeing where the graph crosses the x-axis. Those places are the solutions of the equation. (Consult the manual that came with your calculator to see how to graph functions effectively.)

Before we start, we must emphasize the need to check the calculated solutions to make sure that they truly are solutions by substituting them into the original equation. Sometimes when manipulating an equation to find a solution, we run the risk that the "solutions" we calculated do not satisfy the original equation. Such "solutions" are called **extraneous**. Here are some examples of when extraneous solutions can creep into a problem.

1)   Whenever we square both sides of an equation. For example, if we square both sides of the equation $x = 2$, we get $x^2 = 4$. The second equation has a solution $x = -2$, which clearly is not a solution to the original equation.

2)   When we have denominators, which of course can't be equal to zero. Consider $\dfrac{x-1}{x} = \dfrac{-1}{x}$. Multiplying both sides of this equation by x gives the equation $x - 1 = -1$, which has the solution $x = 0$. However, this solution does not satisfy the original equation.

3)   If the given equation has terms in it like $\sqrt{x + 1}$, $\log(x^2 - 1)$, or $(x - 2)^{\frac{3}{2}}$, we must make sure that our "solution" is a number for which such terms are defined.

4)   In word problems, we must make sure that calculated solutions are physically reasonable. For example, if a word problem leads to the result that the length of a rectangle is -3 or $+5$, we must ignore the solution -3, since it does not make sense to have a negative length.

Let's begin by examining some problems of the onion-peeling variety.

EXAMPLE 1   Solve for x: $\sqrt{x} - 3 = 0$.

Solution This is not of degree 1, but it is of "onion type" because x is in one place only. To get at it, we remove the outermost layer by adding 3 to both sides to get:

$$\sqrt{x} = 3.$$

Next, we remove the square root by squaring both sides to get

$$\left(\sqrt{x}\right)^2 = 3^2$$

or $x = 9.$

(Don't forget to check your solution!)   ∎

EXAMPLE 2   Solve for x: $\left(\sqrt{x} + 2\right)^3 - 64 = 0$.

Solution As in the last example, this equation is not degree 1 but it is of "onion type" because x is in one place only. To get at it, remove the outermost layer by adding 64 to both sides:

$$\left(\sqrt{x} + 2\right)^3 = 64.$$

Next, remove the cube by taking the cube root of both sides of the equation,

$$\left(\left(\sqrt{x} + 2\right)^3\right)^{1/3} = 64^{1/3}$$

or $\sqrt{x} + 2 = 4.$

Subtract 2 from both sides to give

$$\sqrt{x} = 2,$$

and then square both sides to get

$$x = 4.$$

Again, remember to check your solution!   ■

EXAMPLE 3   Solve for x: $\dfrac{1}{x - 5} + \dfrac{1}{x + 5} = \dfrac{10}{x^2 - 25}.$

Solution This is not of onion type, but we can change that by multiplying **both sides of the equation** by $x^2 - 25$ to get rid of the denominators. Doing so we get

$$\frac{x^2 - 25}{x - 5} + \frac{x^2 - 25}{x + 5} = 10.$$

By factoring and canceling we get

$$x + 5 + x - 5 = 10,$$

or $2x = 10$

and $x = 5.$

But $x = 5$ cannot be used in the original equation without dividing by zero. Hence there is no solution to this problem. Good thing we checked.   ■

Sometimes an equation can be reduced to a quadratic equation with a simple substitution. If an equation in x has three terms, one constant, one like x^m, and the other like x^{2m}, then the substitution $y = x^m$ leads to a quadratic equation in y. Consider the following example.

EXAMPLE 4   Solve for x: $x^4 - 5x^2 + 4 = 0$.

Solution This is of quadratic type because if we use the substitution $y = x^2$, then this equation becomes

$$y^2 - 5y + 4 = 0,$$

which is quadratic and has roots $y = 1$ and $y = 4$. (Check this!) But 1 and 4 are NOT the solutions of the given equation. Instead, we have $x^2 = 1$ or $x^2 = 4$, giving us the solutions $x = \pm 1$ and $x = \pm 2$.   ■

We have to be very careful, however, since taking square roots may give us solutions that are not real-valued. Consider the following example.

EXAMPLE 5   Solve for x: $x^4 - 5x^2 - 36 = 0$.

Solution Again, letting $y = x^2$, we will obtain the quadratic equation

$$y^2 - 5y - 36 = 0,$$

which has roots $y = -4$ and $y = 9$. Since $y = x^2$, we have $x^2 = -4$ and $x^2 = 9$, giving $x = \pm 2i$ and $x = \pm 3$.

However, if only real solutions are asked for, the solutions are just 3 and -3.   ■

EXAMPLE 6   Solve for real x: $x^6 + 6x^3 - 16 = 0$.

Solution Now we use the substitution $y = x^3$ to obtain:

$$y^2 + 6y - 16 = 0,$$

which has solutions $y = -8$ and $y = 2$.

So: $x^3 = -8$
and $x = -2,$

or: $x^3 = 2$
and $x = \sqrt[3]{2} = 2^{1/3}$.   ■

EXAMPLE 7   Solve for real x: $x + \sqrt{x} - 6 = 0$.

Solution This is of quadratic type, since if we let $w = \sqrt{x}$ we get $w^2 = x$, and so

$$w^2 + w - 6 = 0.$$

The solutions of this quadratic are $w = 2$ or -3. (Do you agree?) Therefore $\sqrt{x} = 2$ or -3, where the -3 is obviously an extraneous solution. Hence $\sqrt{x} = 2$ or $x = 4$. Check it!   ■

For more complicated equations with coefficients that are also variables, the method is the same, but you should always verify (that is, check) the solutions.

The Zero-Factor Property (ZFP)

If a and b are real numbers, and $a \cdot b = 0$, then either $a = 0$, or $b = 0$, or both. (This result is used to solve equations where there are products of factors.) Consider the following.

EXAMPLE 8   Solve for x: $(x - 1)(x + 2) = 0$.

Solution Here we can use the Zero-Factor Property. By associating $x - 1$ with a and $x + 2$ with b we see that this equation is equivalent to $a \cdot b = 0$. Applying the ZFP we deduce that either $a = 0$ or $b = 0$ (or both), and so

$$x - 1 = 0$$

or

$$x + 2 = 0.$$

The solution to the first equation is clearly $x = 1$, and the solution to the second equation is $x = -2$. Hence the solutions to the original equation are $x = 1$ or $x = -2$. (Check this!)   ■

Notice that if we multiply the factors in the original equation in this example we get the quadratic equation $x^2 + x - 2 = 0$. Using the quadratic formula from the last section, we of course obtain the same solutions $x = 1$ or $x = -2$.

EXAMPLE 9   Solve for x: $(x^2 - 4)(x^2 + 2x - 3) = 0$.

Solution Applying the ZFP gives the two equations

$$x^2 - 4 = 0$$

or

$$x^2 + 2x - 3 = 0.$$

Each of these equations is quadratic and can be solved using the quadratic formula. The solutions to the first equation are $x = 2$ or $x = -2$. (Make sure you verify this for yourself!)

The solutions to the second equation are $x = -3$ or $x = 1$. (Check this also!) So the solutions to the original equation are: $x = 2$, $x = -2$, $x = -3$, or $x = 1$.   ■

The ZFP can be extended to cases when there are more than two factors. For example, if $a \cdot b \cdot c = 0$, then either $a = 0$, $b = 0$, or $c = 0$. When you have a polynomial equation $f(x) = 0$, the idea is to factor $f(x)$ as far as possible, and then to apply the ZFP. Remember that the right-hand side **must be zero** for the ZFP to apply. If your factoring skills are a little rusty, you can buff them up in Chapter 5.

EXAMPLE 10   Solve for x: $x^5 + 3x^4 - 4x^3 = 0$.

Solution We can factor this by first taking out a common factor, to get

$$x^3(x^2 + 3x - 4) = 0.$$

Next, the second factor can be factored further to give

$$x^3(x + 4)(x - 1) = 0.$$

The ZFP says that $x^3 = 0,$

$$x + 4 = 0,$$

or $x - 1 = 0.$

Hence $x = 0, -4$, or 1. Remember to check the solutions. Again, factoring methods are given in Chapter 5. ■

3.3 Exercises

Find the solutions, both real and complex, for Exercises 1–21. Check your answers.

1) $(x - 3)(x + 6) = 0$

2) $(x + 2)(x - 9) = 0$

3) $(2x + 1)(x - 2) = 0$

4) $(2x + 3)(3x - 2) = 0$

5) $(x^2 - 9)(x + 6) = 0$

6) $(x - 1)(x + 2)(x - 3) = 0$

7) $(x^2 - 5)(x^2 - 16) = 0$

8) $(x^2 - 1)(x^2 + 1) = 0$

9) $(x^2 + 1)\left(\dfrac{x}{3} - 4\right) = 0$

10) $(x^2 + 4x - 5)(3x^2 - 81) = 0$

11) $z^4 - 8z^2 + 16 = 0$

12) $s^4 - 5s^2 + 4 = 0$

13) $x^4 - 64 = 0$

14) $(x^2 - 4x - 5)(3x^2 + 81) = 0$

15) $s^4 - 9s^2 + 20 = 0$

16) $\sqrt{x} - 4 = 0$

17) $\sqrt{x} - 3 = 5 - \sqrt{x}$

18) $\left(\sqrt{2x} - 3\right)^3 + 1 = 0$

19) $\left(\dfrac{1}{\sqrt{x}} - 3\right)^4 - 1 = 0$

20) $\dfrac{1}{\sqrt[3]{x^4}} = \dfrac{1}{\sqrt[3]{x}}$

21) $\dfrac{4}{\sqrt{x^3}} = \dfrac{1}{\sqrt{x}}$

22) Find all real solutions to $x^{2/5} - 3x^{1/5} + 2 = 0$.

23) Find all solutions to $\dfrac{1}{x + 1} + \dfrac{1}{x} = \dfrac{3}{x^2 + x}$.

(*Hint:* $x^2 + x$ is a common denominator for the two algebraic fractions on the LHS.)

24) Find all real solutions to $\dfrac{1}{x-6} + \dfrac{1}{x+6} = \dfrac{4}{x^2-36}$.

25) Find all solutions to $\dfrac{1}{x-1} + \dfrac{1}{x+1} = \dfrac{2}{x^2-1}$.

26) Find all real solutions to $\dfrac{1}{x-4} + \dfrac{1}{x+4} = \dfrac{4}{x^2-16}$.

27) Find all real solutions to $x^6 - 26x^3 - 27 = 0$.

28) Find all real solutions to $\dfrac{3}{x} - \dfrac{3}{x+2} = 2$.

29) Find all real solutions to $\dfrac{2}{x} - x = 5$.

30) Let R_1 and R_2 represent the resistances (in ohms) of the two resistors shown in the diagram. The effective resistance R of the parallel circuit shown is given by the equation:

$$\frac{1}{R} = \frac{1}{R_1} + \frac{1}{R_2}.$$

If $R = 7$ ohms and $R_1 = 10$ ohms, then what is the value of R_2?

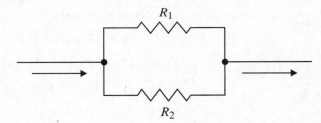

Functions and Their Graphs

4.1 Introduction

In calculus, we are concerned with quantities whose values are not constant, but rather depend on the value of some other quantity, often time. For example, the temperature displayed in Times Square in New York City changes quite a bit, but it has a particular value at any given point in time. We say the temperature is a **function** of time.

EXAMPLE 1   Here is the recording of the temperature in Times Square, N.Y.C., between 8 AM and 8 PM on a certain day:

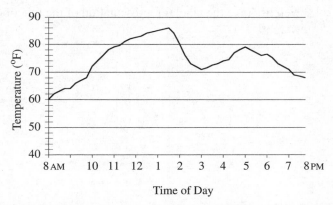

Time of Day

a) Approximately what was the temperature at noon? b) What was the highest temperature, and when did it occur? c) What was the lowest, and when did it occur? d) Give a rough estimate of the average temperature over that 12-hour interval. e) What possible explanations could there be for the dip in the temperature in midafternoon?

Solution a) At noon, the temperature was roughly 83°F.

      b) The highest temperature was about 86°F, at 1:30 PM.

      c) The lowest temperature was about 60°F, at 8 AM.

      d) The average was obviously less than 80°F and greater than 70°F, maybe around 75°F.

e) Here are some possibilities: a passing shower, shade from a tall building, King Kong blocking the sunlight. ▪

The graph shown in the previous example is sometimes called a **line graph,**
even though it is obviously not a line. Quantities that vary as a function of another
quantity can always be represented by such a graph. Another way that a function can
be represented is in the form of a table, and in using such a representation it is important to know how to translate the data into a graph.

EXAMPLE 2   We're off to the post office to mail a letter. We check the cost of mailing a letter first-
class and find this table:

Weight Not Over (oz.)	Cost
1	$ 0.33
2	0.55
3	0.77
4	0.99
5	1.21
6	1.43

Here, the cost of mailing a first-class letter is a function of how much the letter
weighs. Take the tabular representation of the function and graph cost as a function of
weight.

Solution

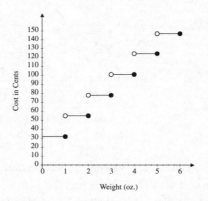

Notice that the open circles indicate missing points, while the solid circles indicate
included points. For example, if the weight is exactly 1 ounce, the cost is 33 cents, not
55 cents. Also notice that this is a line graph, although a strange one. The function is
said to be **piecewise defined.** ▪

So we see that functions can be defined by tables and by graphs. They can also
be defined by words and by symbolic expressions. This fact is sometimes referred to
as the *Rule of Four*. It is important that you can easily change functional representa-
tions from any one form to another. Consider the next example.

EXAMPLE 3   Hurts Car Rental charges $50 for a weekend, plus 15 cents per mile driven. Let C be the rental cost, in dollars, and let d be the distance driven, in miles. a) Express C symbolically as a function of d. b) Graph C as a function of d.

Solution   a) $C = 50 + (\text{number of miles driven})(0.15)$
$C = 50 + (0.15)(d)$

b)   To find the graph of C as a function of d, it is useful to make a table of values of d and the corresponding values of C, like the following:

d (in miles)	C (in dollars)
0	50
50	57.50
100	65
200	80
300	95
500	125
1000	200

Each line of the table yields a point in the graph. For example, the first line goes to the point (0, 50) in the d-C plane. Placing the point into the plane is called **plotting** the point. By plotting all of the points in this table we obtain the following graph:

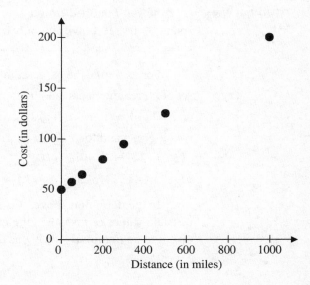

This gives you an incomplete graph, of course, because what if you drive 700 miles or 2000 miles? Well, you can plot those points too. Obviously you'll never finish the complete graph by only plotting points. Can you just draw a smooth curve through the points to get the complete graph? In this case, yes, but in some other cases, no. One of the strengths of calculus is that it tells you whether drawing such a curve to get the complete graph is justified in any given case. In any event, using calculus we can show that in this case it is justified, and so we get the following:

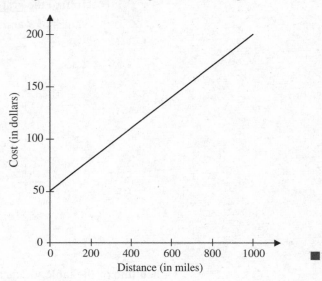

Let's get a little more formal.

DEFINITION
> A **function** f defined from a set A to a set B is a rule that associates with each element of the set A one, and only one, element of the set B.

For example, the expression $x^2 + 2x - 1$ represents a function defined on the set of all real numbers, and it associates the x value of 5 with the $x^2 + 2x - 1$ value of $25 + 10 - 1 = 34$. It also pairs the x value of 7 with the number 62, and so on. If we let $g(x)$ stand for $x^2 + 2x - 1$, that is if $g(x) = x^2 + 2x - 1$, we can write $g(5) = 34$, $g(7) = 62$, $g(0) = -1$, $g(\sqrt{2}) = 2 + 2\sqrt{2} - 1 = 1 + 2\sqrt{2}$, $g(-1) = 12221522$, $g(a + b) = (a + b)^2 + 2(a + b) - 1$, etc.

In the definition above, A is called the **domain** of the function, and the set of all possible values of f is called the **range** of f. If the domain is not specified, we always take it to be **maximal**—as big as possible.

EXAMPLE 4 Suppose $f(x) = 2x^2$; what is $f(4), f(-3), f(4 + h), f(x + \Delta x)$, and $f\left(\sqrt{\dfrac{x}{2}}\right)$?

(Think of Δx as a variable that happens to look a little different. Get used to it!)

Solution Simply replace x in the expression by the appropriate number. So

$$f(4) = 2 \cdot 4^2 = 2 \cdot 16 = 32,$$

$$f(-3) = 2 \cdot (-32^2) = 2 \cdot 9 = 18,$$

$$f(4 + h) = 2(4 + h2^2 = 2(16 + 8h + h^2)2 = 32 + 16h + 2h^2,$$

$$f(x + \Delta x) = 2(x + \Delta x2^2 = 2(x^2 + 2x \Delta x + (\Delta x2^2)2)$$

$$= 2x^2 + 4x \Delta x + 2(\Delta x2^2)$$

and
$$f\left(\sqrt{\dfrac{x}{2}}\right) = 2\left(\sqrt{\dfrac{x}{2}}\right)^2 = 2 \cdot \dfrac{x}{2} = x. \quad \blacksquare$$

EXAMPLE 5 The **absolute value function** is defined as

$$|x| = \begin{cases} x & \text{if } x \geq 0 \\ -x & \text{if } x < 0. \end{cases}$$

This is an important example of a function that is defined piecewise. It represents the distance on the number line from the number x to the origin. Even though "$-x$" may LOOK negative, it is **NOT** negative if $x < 0$. For example, if $x = -5$, then $-x = -(-5) = 5$.

Also notice that $\sqrt{x^2} = |x|$. (Verify this for $x = 3$, and $x = -3$.) $\blacksquare$

Remark For any function f, we always have the following:
a) $f(x) > 0$ wherever the graph of $f(x)$ is above the x-axis.
b) $f(x) = 0$ wherever the graph of $f(x)$ crosses or touches the x-axis.
c) $f(x) < 0$ wherever the graph of $f(x)$ is below the x-axis.

4.1 Exercises

1) Let $f(x) = x^3 + 2x$. Determine:

a) $f(2)$ b) $f(3)$ c) $f(x + h)$

d) $f(2x)$ e) $f(-x)$ f) $f(2 + \Delta x)$

2) Let $u(t) = \frac{1}{2}t^2 + t$. Determine:

   a) $u(2)$ b) $u(-1)$ c) $u(t + \Delta t)$

   d) $u(-t)$

3) a) Evaluate $|x|$ at $x = 0, \pm 1, \pm 2$, and sketch the graph.

   b) Indicate the parts of the number line where $|x|$ is greater than 2.

   c) Find the domain and range of $|x|$.

4) Evaluate:

   a) $|3 - 4|$ b) $|4 - 3|$ c) $|x^2 - x - 3|$ if $x = 2$.

5) Solve $x^2 - 9 > 0$. (*Hint*: plot points to sketch $x^2 - 9$, and see where the graph is above the x-axis.)

6) Solve $x^2 - 4 < 0$. (*Hint*: plot points to sketch $x^2 - 4$, and see where the graph is below the x-axis.)

7) Let $f(x) = \sqrt{x} + 1$.

   a) What is its domain?   b) Plot a few points to sketch a rough graph.

   c) What is its range?

8) Let $f(x) = \sqrt{x + 2} - 3$.

   a) What is its domain?   b) Plot a few points to sketch a rough graph.

   c) What is its range?

9) Let $g(x) = x^2 + 2$.

   a) Plot a few points to sketch a rough graph.   b) What are its domain and range?

10) It is estimated that nine out of every one million $20 bills are counterfeit. Design both a numerical (tabular) and symbolic representation that gives the predicted number of counterfeit $20 bills in a sample of x million $20 bills for $x = 0, 1, 2, 3, 4, 5, 6$.

11) The following table gives the average precipitation in Reno, Nevada, where $x = 1$ corresponds to January and $x = 12$ to December.

x (month)	1	2	3	4	5	6	7	8	9	10	11	12
P (in.)	0.5	0.3	0.4	0.3	0.2	0.1	0.3	0.5	0.2	0.4	0.4	0.3

   a) Determine the value of P during April and June.

   b) Is P a function of x? Explain.

   c) If $P = 0.3$, determine x.

4.2 Lines and Their Equations

Straight lines are the simplest of all curves, and one of the main ideas of calculus is to use lines to approximate complicated curves. So you've got to be proficient at lines and their equations. Lines have various degrees of steepness, which in mathematics is called **slope.** Consider Figure 4.1.

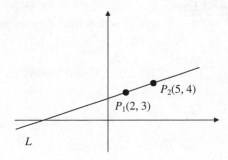

Figure 4.1

As you go from point P_1 to point P_2, you increase vertically a distance of 1 (called the **rise**), and horizontally a distance of 3 (called the **run**). The slope of line L is defined as $\frac{\text{rise}}{\text{run}} = \frac{1}{3}$.

Suppose the line L contains the two points $P_1(x_1, y_1)$ and $P_2(x_2, y_2)$ (see Figure 4.2),

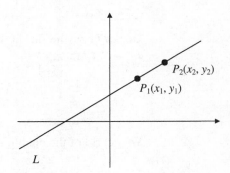

Figure 4.2

then L is said to have slope $m = \dfrac{y_2 - y_1}{x_2 - x_1}$, if $x_2 - x_1 \neq 0$.

Remarks

a) Notice that if $x_2 - x_1 = 0$, then $x_2 = x_1$ and the line is vertical, and hence we have defined the notion of slope only for nonvertical lines.

b) Notice also what happens if we choose not P_1 and P_2, but any two other points on L. By using similarity of triangles we can prove that the slope calculated using any two points on L is equal to the slope obtained using P_1 and P_2.

We have talked about graphs of functions. What about graphs of equations? Here's the idea.

DEFINITION | The **graph of an equation** in x and y is the set of points (a, b) whose coordinates satisfy the equation when a is substituted for x and b is substituted for y.

EXAMPLE 1 Consider the line of slope -1 through the point $(-2, 1)$. Find an equation whose graph is that line. Equivalently: What is the equation of the line of slope -1 through $(-2, 1)$?

Solution Consider this graph:

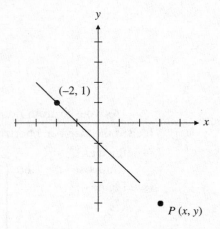

Does $P(x, y)$ lie on the line? It does if, and only if, the slope is correct, that is, if $\frac{\text{rise}}{\text{run}} = -1$, meaning

$$\frac{y - 1}{x - (-2)} = -1.$$

We get rid of the denominator by multiplying by it, and then get

$$y - 1 = (-1)(x + 2).$$

So that is the equation of the line. ■

Theorem

The equation of the line of slope m going through the point (x_0, y_0) is

$$\boxed{y - y_0 = m(x - x_0)}\ .$$

Proof

We'll use the same idea as in the last example.

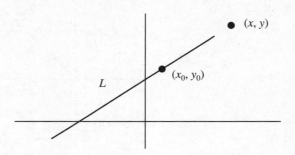

The point (x, y) lies on the line L if, and only if, the slope calculated from (x, y) and (x_0, y_0) is correct, that is: m. In other words $\dfrac{y - y_0}{x - x_0} = m$, which means $y - y_0 = m(x - x_0)$. (Notice that x and y are the variables, while m, x_0, and y_0 are just numbers.) This is called the **point-slope form** of the equation of a line. It is very important! Memorize it! (Right now is an excellent time.)

EXAMPLE 2   Find the equation of the line L of slope -3 going through the point $(2, -5)$.

Solution Nothing to it if you've memorized that equation. In this case, $m = -3$, $(x_0, y_0) = (2, -5)$, and x and y are just the variables in the equation.

So
$$y - (-5) = -3(x - 2),$$
which cleans up to be
$$y + 5 = -3(x - 2).$$

It's easy, once you know the slope and a point on the line.   ■

EXAMPLE 3   Consider the lines through the origin of slope $m = 1, 2, -3$. Find their equations and graph them.

Solution We'll use the form $y - y_0 = m(x - x_0)$. In our case $(x_0, y_0) = (0, 0)$, and so
$$y - 0 = m(x - 0),$$

which simplifies to
$$y = mx.$$

So the three given lines have equations $y = x$, $y = 2x$, and $y = -3x$. Their graphs are shown in Figures 4.3(a), 4.3(b), and 4.3(c) on the following page, respectively. We can get the graphs by plotting merely one point in addition to the origin. Go 1 to the right, and up m, if $m > 0$, or down $|m|$ if $m < 0$.

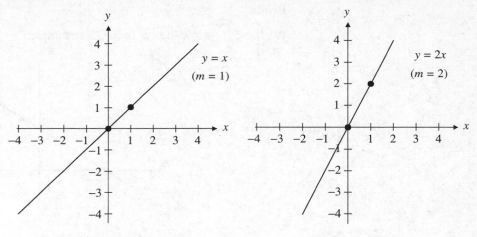

Figure 4.3(a)   Figure 4.3(b)

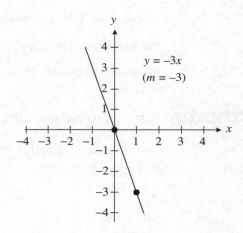

Figure 4.3(c)

Pictured in Figure 4.4 is the **family** of functions of the form $y = mx$, for several m, all on the same set of axes for comparison purposes. As we go from left to right (our usual assumption), if m is positive, the line rises; if it is negative, the line sinks. The larger the absolute value of the slope, $|m|$, the steeper the line becomes. (What happens when the slope is a really big positive number?)

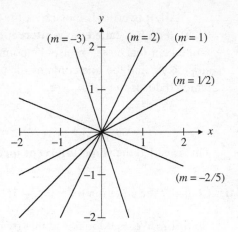

$y = mx$, for several m.

Figure 4.4

Remark It helps, whenever possible, to use the same scale on the x- and y-axes, meaning the distance from 0 to 1 is the same on the two axes. Sometimes that is really not feasible. (See Exercise 5.) In any event, always mark your scales—that is, show where 1, 2, etc. (or other appropriate values) are.

EXAMPLE 4 Graph $h(x) = |x|$.

Solution If $x < 0$, $h(x) = -x$.   If $x \geq 0$, $h(x) = x$.

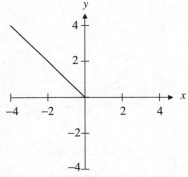 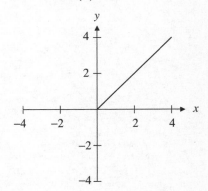

Put it together, and we get $|x|$.

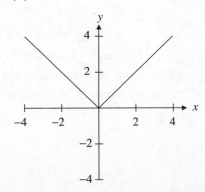

■

All nonvertical lines cross the y-axis. If a line crosses the y-axis at the point $(0, b)$, then b is called the **y-intercept**. To find the equation of the line of slope m and y-intercept b is easy. We use the point-slope form of the equation of a line, where the slope is m, and the line contains the point $(0, b)$. Doing so we get $y - b = m(x - 0)$ which reduces to

$$\boxed{y = mx + b}\ .$$

This is called the **slope-intercept form**. You should also memorize this.

EXAMPLE 5 Consider the equation $6x - 3y = 15$. What is its graph?

Solution We could make a table by putting $x = 1, 2, 3$, etc., and calculating the corresponding y-values, and then plotting points, but we won't. Instead, we solve for y:

$$-3y = -6x + 15$$

or $$y = 2x - 5,$$

which we know is the equation of a line of slope 2 going through $(0, -5)$. If we start from $(0, -5)$ and go over 1 and up m (in this case $m = 2$), we get a second point on the line $(1, -3)$, and hence we know the whole graph.

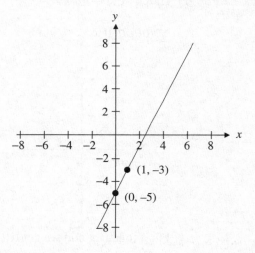

Lastly, we remind you that parallel lines have equal slopes, and perpendicular lines have slopes that are negative reciprocals of each other. This means, if line L_1 has slope m_1, and line L_2 has slope m_2, then, L_1 is parallel to L_2 if, and only if, $m_1 = m_2$; moreover, L_1 is perpendicular to L_2 if, and only if, $m_2 = \frac{-1}{m_1}$. Consider the following.

EXAMPLE 6   Find the equation of the line passing through the point (1, 3) that is parallel to the line $2x + 3y = 6$.

Solution First, we determine the slope of the given line by putting it in slope-intercept form. We solve $2x + 3y = 6$ for y to get:

$$y = -\frac{2}{3}x + 2.$$

Comparing this to $y = mx + b$, we see that the slope of the given line is $-\frac{2}{3}$. Since the point on the desired line is (1, 3), we use the point-slope form to get

$$y - 3 = -\frac{2}{3}(x - 1),$$

which is one form of the answer. Depending on your purpose, you may wish to change it to slope-intercept form, or leave it as is.   ■

EXAMPLE 7   Find the equation of the line passing through the point $(3, -2)$ that is perpendicular to the line $y = 6x + 4$.

Solution The line $y = 6x + 4$ has slope equal to 6. Any line perpendicular to this line will have slope equal to $-\frac{1}{6}$. (This is the negative reciprocal of 6, right?) So the line we want passes through the point $(3, -2)$ and has slope $m = -\frac{1}{6}$. Using the point-slope form we obtain:

$$y - (-2) = -\frac{1}{6}(x - 3)$$

or

$$y = -\frac{1}{6}x - \frac{3}{2}. \quad ■$$

4.2 Exercises

In Exercises 1–5, graph the function by plotting two points each. (Why is it enough to plot only two points?)

1) $f(x) = \frac{1}{3}x$

2) $y = -1.5x$

3) $g(x) = 0.6x$

4) $f(x) = 5x$

5) $y = 1000x$

6) Find the equation of the line of slope $\frac{2}{3}$ through the point (2, 7).

7) Find the equation of the line of slope -6 through the point $(-3, 2)$.

8) Find the equation of the line containing the two points $(1, 2)$ and $(-2, 1)$. (*Hint*: Find m.)

9) Find the equation of the line containing the two points $(-1, 4)$ and $(2, 1)$.

10) Graph the line given by the equation $4y - 3x = 24$.

11) Find the point-slope form for the equation of the line through the point $(1, 4)$ parallel to the line $y = 2x - 5$. Graph the line.

12) Find the point-slope form for the equation of the line L through the point $(4, 2)$ parallel to the line $y = 9x - 5$. Graph the line L.

13) Find the slope-intercept form for the equation of the line through the point $(-1, 4)$ perpendicular to the line $y = \frac{1}{3}x - 5$. Graph this line, and label the y-intercept.

14) Find the slope-intercept form of the equation of the line L through the point $(1.34, 2.67)$ perpendicular to the line $y = -0.235x - 5.29$. Graph this line L, and label the y-intercept.

15) Find the equation of the line through the point $\left(0, \frac{4}{3}\right)$ parallel to the line $y = 5x - 10$. Graph both lines.

16) Graph the line given by the equation $4.014y - 2.276x = 9.87$.

17) Find the equation of the line through the point $\left(\frac{1}{2}, -3\right)$ perpendicular to the line $y = -\frac{1}{7}x - 5$. Graph both lines.

18) Find the equation of the line through $(3, -4)$ parallel to the line $2x - y = 5$, and put it into slope-intercept form.

19) The temperature measured in degrees Fahrenheit, F, is a linear function of the temperature measured in degrees Celsius, C. The ordered pair, or point, $(0, 32)$ is an ordered pair of this function because $0°C$ is equivalent to $32°F$, the freezing point of water. The ordered pair $(100, 212)$ is also an ordered pair of this function since $100°C$ is equivalent to $212°F$, the boiling point of water. a) Use these two points to write F as a function of C, and graph it. b) Find the temperature in degrees Fahrenheit of a thermometer reading $45°C$.

20) In statistics, one way to analyze data is by determining the "line of best fit," also called a **regression line**. Once this regression line is calculated, one may use it to predict data values in between the collected data. Given a set of data points $\{(x_1, y_1), (x_2, y_2), (x_3, y_3), \ldots, (x_n, y_n)\}$, the regression line is given by $y = ax + b$, where a and b are functions of the actual data, x_i and y_i, for $i = 1$ to n. (The exact formulas are too complicated to give here right now.) Suppose a particular set of data gives rise to values $a = 1.078$ and $b = 3.4457$. Graph this particular line of regression. What is y if $x = 2.34$? What is the significance of the y-intercept?

4.3 Power Functions

Functions of the form $f(x) = x^r$, where r is a constant, are called **power functions**. Let's look at their graphs.

Part A: $y = x^r$, with $r = 2, 4, 6$, etc. (See Figure 4.5).

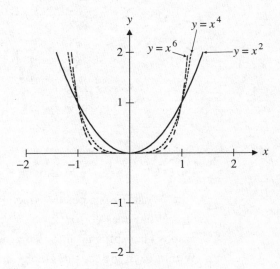

Figure 4.5

Notice how the points $x = 1$ and $x = -1$ are "pivotal"? Consider $x > 0$, or the right side of the graph. For $x < 1$, the relative position of the curves is opposite to what it is when $x > 1$. Well, let's think about it: for $x < 1$, the more we multiply it by itself, the smaller it gets. The higher the power, the smaller the number gets. When $x > 1$, the opposite is true. The higher the power, the more we multiply by a number that is greater than 1, hence the larger the number gets. Notice that these graphs seem symmetric about the y-axis. In fact, they are. It doesn't matter whether we insert x or $-x$ into an even-powered function; the result is the same. (Can you show this?) This shows **symmetry about the y-axis.**

Part B: $y = x^r$, with $r = 3, 5, 7$, etc.

Symmetry through the origin is exhibited by the odd-powered functions. Figure 4.6 shows three members of this family.

As in Figure 4.5, the points $x = 1$ and $x = -1$ are pivotal in that the relative positioning of the graphs changes there. Now, however, the graphs are symmetric through the origin. That is, if the point (x, y) is on the graph, then so is $(-x, -y)$.

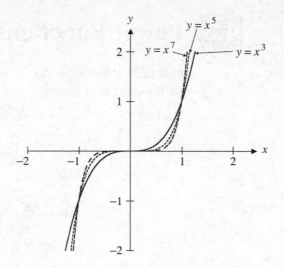

Figure 4.6

Part C: $y = x^r$, with $r = \frac{1}{2}, \frac{1}{4}, \frac{1}{6}$, etc., and $r = \frac{1}{3}, \frac{1}{5}, \frac{1}{7}$, etc.

What if r is a positive fraction of the form $\frac{1}{n}$, where n is a whole number 1, 2, 3, etc.? What does $y = x^{1/2}$, $y = x^{1/3}$, etc. look like? Well, as it turns out, this family of graphs also splits up into two groups, namely, the even n graphs and odd n graphs. Within each group, the graphs display similar behavior. For $y = x^{1/2}$ and $x^{1/4}$, the graphs look like that in Figure 4.7. Since $y = x^{1/n}$ is equivalent to the nth root, if n is even, x must not be negative.

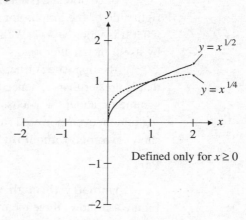

Figure 4.7

The other graphs of $y = x^{1/n}$ for n even are very similar to those shown in Figure 4.7, and as is the case in Figures 4.5 and 4.6, the point $x = 1$ is where the relative positioning of the graphs changes. If n is odd—say, for example, $y = x^{1/3}$ or $y = x^{1/5}$— then x can be any real number. The graphs of these functions are shown in Figure 4.8.

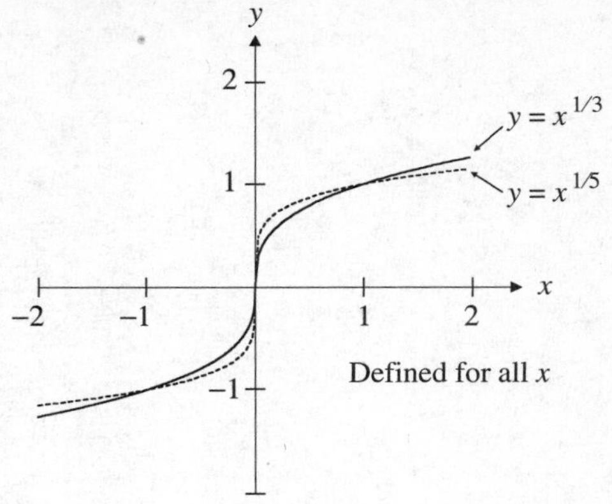

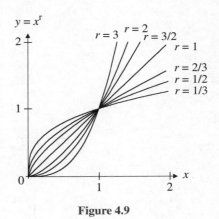

Figure 4.8

The other graphs of $y = x^{1/n}$ for n odd are very similar to those shown in Figure 4.8, with the same ordering change at $x = 1$ and $x = -1$.

To sum up Parts A, B, and C: for $x \geq 0$, $y = x^r$ is shown in Figure 4.9.

Figure 4.9

All of these are examples of **power functions**. Notice that $x^{2/3}$ is calculated by squaring $x^{1/3}$.

Part D: $\qquad y = x^r$, with $r = -1, -3$, etc., and $-2, -4$, etc.

What if r is a negative integer? Can you guess that the situation will divide up into two cases, r even or odd? Figure 4.10 shows the situation for r odd: $\dfrac{1}{x}, \dfrac{1}{x^3}$, etc. (Recall that x^{-3} means $\dfrac{1}{x^3}$, etc.)

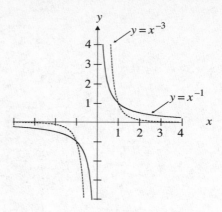

Figure 4.10

For $r = -2$ and -4, the graph is shown in Figure 4.11.

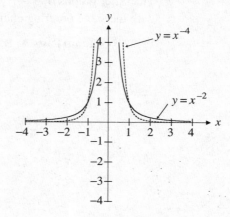

Figure 4.11

The functions x^{-3}, x^{-5}, etc. all basically look like x^{-1}, and x^{-4}, x^{-6}, etc. basically look like x^{-2}. The points $x = 1$ and $x = -1$ are again important since the relative positioning changes there.

4.3 | Exercises

Using graph paper, calculate and plot exactly at least four points each for Exercises 1–9.

1) $f(x) = \sqrt{x}$

2) $f(x) = x^3$

3) $g(x) = \sqrt[3]{x}$

4) $f(x) = \dfrac{1}{x}$

5) $g(x) = \dfrac{1}{x^2}$

6) $w(x) = x^{-3}$

7) $f(x) = x^{2/3}$ 8) $f(x) = x^{3/2}$

9) $f(x) = \dfrac{1}{\sqrt{x}}$

10) Using graph paper, calculate and plot at least six points for $f(x) = x^2$. Now sketch a curve through these points as an approximation to the graph of f. Similarly, on the same graph paper, sketch $g_1(x) = 2x^2$, $g_2(x) = \frac{1}{2}x^2$, $g_3(x) = -2x^2$, and $g_4(x) = -\frac{1}{2}x^2$. How are these four functions related to the original function f?

11) Sketch $y = ax^n$, where n is an integer greater than 2 and a is real. Consider the cases for n even or n odd separately, and plot the graphs for $a > 0$ or $a < 0$ on the same axes.

4.4 Shifting Up or Down

How does the graph of $y = x^2 + 2$ compare to the graph of $y = x^2$? Well, all the y-coordinates of the first graph are 2 bigger than those of the second graph. But y is the altitude, or height, of the point (x, y). So to go from the graph of $y = x^2$ to $y = x^2 + 2$, just push up the graph a distance of 2, as shown in Figure 4.12.

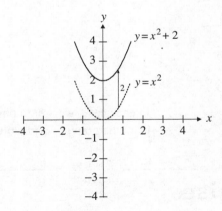

Figure 4.12

Also consider the examples in Figures 4.13 and 4.14.

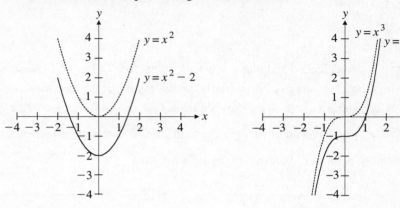

Figure 4.13(a)   Figure 4.13(b)

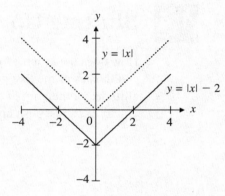

Figure 4.14

In each case, we are graphing a function by using the fact that it is shifted up or down from the graph of a function that we know.

4.4 Exercises

1) Graph the following functions:

   a) $y = \dfrac{1}{2}x$

   b) $y = \dfrac{1}{2}x + 2$

   c) $y = \dfrac{1}{2}x - 1$

2) Graph the following functions:

   a) $y = \dfrac{1}{2}x^2$

   b) $y = \dfrac{1}{2}x^2 + 1$

   c) $y = \dfrac{1}{2}x^2 - 2$

3) Graph the function $y = -2x + 3$.

4) Graph the function $y = x^3 + 5$.

5) Graph the following functions:

   a) $y = x^2 - 4$ b) $y = x^2 + \pi$ c) $y = x^{-2} + 1$

6) Graph the following functions:

   a) $y = x^2 + 2$ b) $y = x^{-2} - 2$

7) Graph the function $y = \sqrt{x} + 2$ for $x \geq 0$.

8) Graph $|x|$ and $|x| + 1$.

9) Let $f(x) = \dfrac{1}{x}$. Graph $f(x)$ and $f(x) + 1$.

10) Let $g(x) = \dfrac{1}{x^2}$. Graph $-g(x)$ and $2 - g(x)$.

4.5 Shifting Left or Right

Consider the three graphs in Figure 4.15.

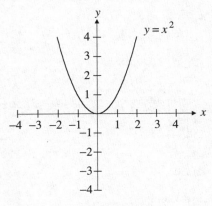

Figure 4.15(a)

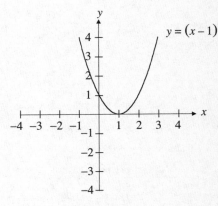

Figure 4.15(b)

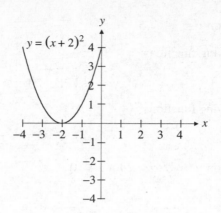

Figure 4.15(c)

(Don't take **our** word for it; test these by plotting points for $x = 0, \pm 1, \pm 2, \pm 3$.)

Notice that going from $f(x) = x^2$ to $f(x - 1) = (x - 1)^2$ we needed to shift the graph to the **right** a distance of 1, and going from $f(x) = x^2$ to $f(x + 2) = (x + 2)^2$ shifted the graph to the **left** a distance of 2. **This is true for all functions $f(x)$**, not just $f(x) = x^2$. In general, if we suppose a is some positive number, then the graph of $f(x - a)$ is the graph of $f(x)$ moved to the right a distance of a, and the graph of $f(x + a)$ is the graph of $f(x)$ moved to the left a distance of a. For instance, check the graphs in Figure 4.16.

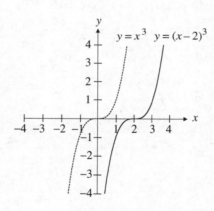

Figure 4.16(a)

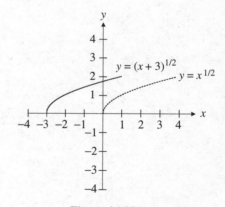

Figure 4.16(b)

Remarks

a) Notice in Figures 4.15(c) and 4.15(b), that the "vertex" of $(x + 2)^2$ is at -2, while the vertex of $(x - 1)^2$ is at $+1$, which may be opposite to what you might have expected, but plotting a few points proves it.

b) It is important to see that, for example, $(x + 3)^2$ is very different from $x^2 + 3$. If we start from the graph of x^2, $(x + 3)^2$ moves it **3 units to the left,** while $x^2 + 3$ moves it **up a distance of 3 units.**

4.5 Exercises

1) Graph x^2, $(x + 1)^2$, and $(x - 1)^2$, by plotting five points for each.

2) Graph the following functions:

   a) $f(x) = (x - 3)^2$

   b) $y = (x + \pi)^2$

   c) $y = \dfrac{1}{x - 1}$

3) Graph the following functions:

   a) $y = (x + 3)^3$

   b) $y = \sqrt[3]{x + 1}$

   c) $y = -\dfrac{1}{(x - a)^4}$, where $a > 0$

4) Graph the function $y = \sqrt{x - 4}$ for $x \geq 4$.

5) Graph $g(x) = |x|$, and $h(x) = |x - 3|$.

6) Graph the function $y = \sqrt{x + 2}$ for $x \geq -2$.

7) Let $B(t)$ be the number of births during year t in the United States. Assume for the moment that 60% of these people will graduate from high school at the age of 18. Let $G(t)$ be the number of graduates from high school during year t in the United States. What approximate relation would you expect between the functions $B(t)$ and $G(t)$? How could this be used by college admissions staff members?

4.6 Translations

We can put all this shifting together to get functions all over the place. Such combinations of vertical and horizontal shifts are called **translations**.

EXAMPLE 1   Graph the function $y = \dfrac{1}{(x - 1)} + 2$.

Solution We graph this by a) starting with the graph for $y = \frac{1}{x}$, b) shifting this right 1 unit to get $y = \frac{1}{(x - 1)}$, and then c) shifting up 2 to get the final graph $y = \frac{1}{(x - 1)} + 2$.

a)

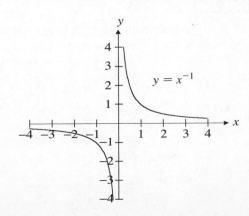

$y = x^{-1}$

b)

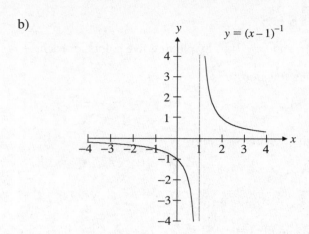

Notice that the vertical asymptote at $x = 0$ moves over to $x = 1$. (An **asymptote** is a straight line that a curve approaches closer and closer. The rigorous definition will be given in your calculus book, because it involves the notion of limit.)

c)

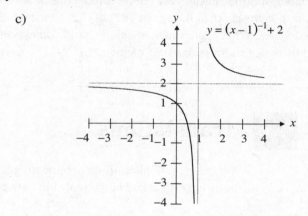

Notice that the horizontal asymptote at $y = 0$ moves up to $y = 2$.   ■

EXAMPLE 2   Graph $g(x) = x^2 - 6x + 5$.

Solution The trick is to complete the square.

$$g(x) = (x^2 - 6x + 9) + (5 - 9)$$
$$= (x - 3)^2 - 4$$

So, $g(x)$ looks like x^2 but translated 3 units right and 4 units down.

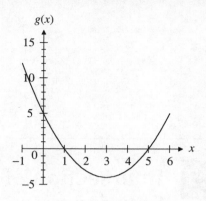

4.6 Exercises

Graph the following functions:

1) a) $y = x^2$ b) $y = x^2 - 1$

   c) $y = \frac{1}{2}(x^2 - 1)$ d) $y = \frac{1}{2}(x^2 - 1) + 2$

2) a) $y = x^2$ b) $y = x^2 - 3$

   c) $y = (x - 2)^2 - 3$

3) a) $y = \frac{1}{x}$ b) $y = \frac{1}{x - 2}$

   c) $y = \frac{3}{x - 2}$ d) $y = \frac{3}{x - 2} - 1$

4) a) $y = \frac{1}{x^2}$ b) $y = \frac{1}{(x + 1)^2}$

   c) $y = \frac{1}{(x + 1)^2} + 4$

5) a) $y = \sqrt{x}$ b) $y = \sqrt{-x}$

   c) $y = \sqrt{-(x - 2)} = \sqrt{2 - x}$ d) $y = 1 + \sqrt{2 - x}$

6) $y = (x - 1)^2 - 1$

7) $y = \frac{1}{x + 1} + 2$

8) $y = (x - 2.34)^3 - 7.21$

9) $y = 2 - \sqrt{x - 2}$

10) $y = \dfrac{1}{x - 1} + 5$

11) $y = (x - 1)^3 + 2$

12) $y = 1 - \dfrac{1}{(x - \pi)^4}$

4.7 Intersection of Curves and Simultaneous Solutions

We know that a point (a, b) lies on a curve in the xy–plane if, and only if, $x = a$ and $y = b$ are solutions of the equation of that curve. (For example, the curve $y = x^2$ has the point $(2, 4)$ on it because $2^2 = 4$.) What about a point of intersection of two curves? Well, that point lies on both curves, so its coordinates must satisfy the equations of both curves at the same time! That is, x and y must satisfy the two equations **simultaneously.** A set of two (or more) equations of two (or more) variables is called a **system of equations.** Finding solutions of such systems of equations is usually easy.

EXAMPLE 1   Find the intersection points of the curves whose equations are $y = x^2 - 4$ and $x + y = 8$.

Solution Since both equations have to be true at the same time, we can solve for either x or y in one equation and substitute that result into the other equation. Here, let's solve for y in the first equation (it's already done!) and substitute that into the second equation, to give us

$$x + (x^2 - 4) = 8.$$

We clean this up and put it into standard form:

$$x^2 + x - 12 = 0.$$

Here we can either use the quadratic formula or factor it to get

$$(x + 4)(x - 3) = 0.$$

By the Zero-Factor Property, $x = 3$ or -4. (See Section 3.3 for more on the ZFP.) So we know that there are two points of intersection: one where $x = 3$, and the other where $x = -4$. To find the corresponding y-values, you can substitute each x-value into either of the two equations (pick the easier one) to calculate y.

So: $y = x^2 - 4$

hence if $x = 3$, $y = 3^2 - 4 = 5$, and if $x = -4$, $y = (-4)^2 - 4 = 12$. Therefore, the two points of intersection are $(3, 5)$ and $(-4, 12)$. ∎

EXAMPLE 2 Where do the lines

$$2x + 3y = 7$$

and

$$-3x + y = 11$$

intersect?

Solution Here, let's start with the second equation to get $y = 3x + 11$, and substitute that into the first equation to get

$$2x + 3(3x + 11) = 7.$$

We clean this up and solve:

$$2x + 9x + 33 = 7$$

$$11x = -26$$

or

$$x = \frac{-26}{11} = -\frac{26}{11}.$$

To find the corresponding y-value, use $y = 3x + 11$ with $x = -\frac{26}{11}$ in it to get

$$y = 3\left(-\frac{26}{11}\right) + 11$$

or

$$y = -\frac{78}{11} + 11 = \frac{-78 + 121}{11} = \frac{43}{11}.$$

Hence there is one intersection point: $\left(-\frac{26}{11}, \frac{43}{11}\right)$. ∎

Remark The method we used in these two examples is called the **method of substitution.** If all the equations are of degree 1 in whatever variables there are (as in Example 2), there is another way we can write the system using matrices. You can then use methods from the mathematical subject of linear algebra to easily solve such problems.

4.7 Exercises

1) Find the simultaneous solutions to the following systems of equations:

a) $\begin{cases} 3x - 2y = 16 \\ 5x + y = 5 \end{cases}$ b) $\begin{cases} x^2 - 4y = 6 \\ 2x + 2y = 3 \end{cases}$

2) Find the intersection of the curves:

a) $x^2 = y - 2$ and $2x + 3y = -7$ b) $x + y = 2$ and $\sqrt{x} + 4y = -6$

3) Find the points of intersection for the line $y = x$ and the circle $x^2 + y^2 = 1$. (*Hint*: Graph the functions.)

4) Find the intersection of the curves:

a) $x^2 = 2.20y - 1$ and $2.01x + 3.2y = -2$

5) Find all the intersection points of the unit circle centered at the origin with the line $x = \frac{1}{2}$. Sketch.

6) Find the points of intersection for the line $y = -x$ and the circle $x^2 + y^2 = 1$.
(*Hint*: Graph the functions.)

7) Sketch the circle of radius 2 centered at $(1, 1)$. Find all points on this circle whose altitude (height above the x-axis) is 2.

Changing the Form of a Function

Most limit problems that you'll meet will include a fractional expression in which the numerator and denominator have a common factor that needs to be canceled out. Moral: you need to know how to factor to be a player in this game. It isn't always obvious how you should factor, but in your factoring toolbox you've got four basic methods that should be tried in this order:

1) Common factors

2) Special formulas

3) Grouping

4) The Factor Theorem

Recall that factoring an expression means writing it as a product.

5.1 Common Factors

The easiest tool to use is to take out all factors that are common to all the terms of the expression.

EXAMPLE 1 Factor $3x^2y^3 + 15xy^4 - 21x^3y^2$.

Solution Notice that all three terms share a factor of 3, as well as x and y^2. So, taking out this common factor of $3xy^2$, we get

$$3x^2y^3 + 15xy^4 - 21x^3y^2 = 3xy^2(xy + 5y^2 - 7x^2). \quad \blacksquare$$

Remark This is the easiest method of factoring and should always be done before going on to the rest.

5.1 Exercises

Factor the following expressions:

1) $2xy + 4x$

2) $6wz + 2wzt$

3) $xy + 4x + 2xw$

4) $6x^2y + 3xy + 9xy^2$

5) $10x^8y^6 + 25x^2y^4 + 20x^3y^{10}$

6) $24x^2yz + 2xy^2z^2 + 4xyz^3$

5.2 Special Formulas

1) $x^2 - y^2 = (x + y) \cdot (x - y)$
2) $x^3 + y^3 = (x + y) \cdot (x^2 - xy + y^2)$
3) $x^3 - y^3 = (x - y) \cdot (x^2 + xy + y^2)$
4) $x^2 + (a + b)x + ab = (x + a) \cdot (x + b)$
5) $acx^2 + (bc + ad)x + bd = (ax + b) \cdot (cx + d)$
6) $x^2 + 2xy + y^2 = (x + y)^2$ and $x^2 - 2xy + y^2 = (x - y)^2$

The formulas of 6 are special cases of the two previous formulas. They are called **perfect squares**.

Remark You should know formulas 1, 2, 3, and 4 "in your sleep." The others, especially 5, are not so critical. All of them can be checked by multiplying out the right-hand sides.

Formula 1 is called the **difference of two squares** and comes up a lot. Consider the following examples.

EXAMPLE 1 a) $z^2 - 9 = (z + 3) \cdot (z - 3)$

Here we are using 1 with $x = z$ and $y = 3$.

b) $x^4 - y^2 = (x^2)^2 - y^2 = (x^2 + y) \cdot (x^2 - y)$

Here x is replaced by x^2.

c) $(x - y)^2 - 4y^2 = (x - y + 2y) \cdot (x - y - 2y)$

$$= (x + y) \cdot (x - 3y) \blacksquare$$

Formula 2 is called the **sum of two cubes**, while formula 3 is the **difference of two cubes**. Notice how similar they are. One way to remember the signs in a formula is as follows: whatever sign is between the two cubes matches the sign between x and y in the first factor of the product, while the sign in front of the xy in the second factor is opposite to that. (Notice the $-xy$ in the middle of formula 2 and the xy in the middle of formula 3. It sometimes surprises people.)

EXAMPLE 2 a) $a^3 + 8b^3 = a^3 + (2b)^3 = (a + 2b) \cdot (a^2 - 2ab + 4b^2)$

b) $27x^3 + 64y^3z^6 = (3x)^3 + (4yz^2)^3$
$$= (3x + 4yz^2) \cdot (9x^2 - 12xyz^2 + 16y^2z^4)$$

Here we are using formula 2 with x replaced by $3x$ and y replaced by $4yz^2$.

c) $a^3 + 2b^3$ is the sum of the cubes of a and $\sqrt[3]{2}b$ (which equals $2^{1/3}b$).. So
$$a^3 + 2b^3 = (a + 2^{1/3}b) \cdot (a^2 - 2^{1/3}ab + 2^{2/3}b^2). \quad \blacksquare$$

The usage of formula 3 goes pretty much the same way as that of 2.

EXAMPLE 3 $x^3 - 64y^6 = (x - 4y^2) \cdot (x^2 + 4xy^2 + 16y^4)$

Again watch the sign of that middle term in the last factor! $\blacksquare$

Formula 4 is used by noticing that the middle term is the **sum** of a and b, while the last term is their **product**.

EXAMPLE 4 Factor $x^2 + 5x + 6$.

Solution We are looking for two numbers a and b so that their product is 6 and their sum is 5. Well, 6 factors only as 2 times 3, 6 times 1, -2 times -3, or -6 times -1. Of all these possibilities the only pair to add up to 5 is 2 and 3.

So: $x^2 + 5x + 6 = (x + 2) \cdot (x + 3). \quad \blacksquare$

EXAMPLE 5 Factor $y^2 - 10y + 21$.

Solution We need two numbers whose product is 21, so we think of 3 and 7, -3 and -7, 21 and 1, and -21 and -1. However, we notice that their sum must be -10, so we use -3 and -7.

$$y^2 - 10y + 21 = (y - 3) \cdot (y - 7). \quad \blacksquare$$

EXAMPLE 6 Factor $s^2 + 2s - 8$.

Solution Now we need two numbers whose product is -8 and whose sum is 2. The number -8 is the product of -8 and 1, -4 and 2, 4 and -2, or 8 and -1. The correct pair that adds up to 2 is 4 and -2, hence

$$s^2 + 2s - 8 = (s + 4) \cdot (s - 2). \quad \blacksquare$$

EXAMPLE 7   Factor $a^4 - a^2 - 6$.

Solution This looks different, but if we let $x = a^2$, we get $x^2 - x - 6$, and then this method of factoring applies. Now, -6 is the product of -6 and 1, -3 and 2, -2 and 3, and -1 and 6. Since their sum has to be -1, the only possible choice is -3 and 2. So,

$$x^2 - x - 6 = (x - 3) \cdot (x + 2)$$

and hence

$$a^4 - a^2 - 6 = (a^2 - 3) \cdot (a^2 + 2).$$

This can be factored further to give

$$a^4 - a^2 - 6 = (a - \sqrt{3})(a + \sqrt{3})(a^2 + 2). \quad \blacksquare$$

Remark   In the last example the factorization to $(a^2 - 3) \cdot (a^2 + 2)$ is as far as you can go if you factor "over the integers," meaning that all the coefficients must be integers. If you factor "over the reals," you must go the extra step and get $(a - \sqrt{3})(a + \sqrt{3})(a^2 + 2)$. In some other contexts, you can even factor $a^2 + 2$ "over the complex numbers" into $(a + \sqrt{2}i)(a - \sqrt{2}i)$. In most situations in calculus, you'll need to factor only over the reals.

EXAMPLE 8   Factor $y^2 + 10y - 24$.

Solution Notice how -24 is the product of many pairs, so this method of factoring can get to be tedious—in fact it can be a real pain! Since their sum is 10, the pair that works is 12 and -2, so

$$y^2 + 10y - 24 = (y - 2) \cdot (y + 12). \quad \blacksquare$$

Ray of Hope: If this method is too tedious, or too tough, or if the factors are not whole numbers (which happens often), you can use a different method based on the quadratic formula and the Factor Theorem (F.T.). For example, $x^2 - 2x - 4$ can't be factored easily without the Factor Theorem, which we will see in Section 5.4.

Formula 5 is usually really tough to use because too many possibilities need to be checked, unless you're lucky and hit the right one quickly. Your best bet in this case is to factor out the coefficient of the first term and work with the remaining expression. And there's always the Factor Theorem.

5.2 Exercises

Factor the following expressions. (Here, to factor means to factor as far as possible over the reals; see Example 7.)

1) $4y^2 - 9z^2$

2) $16x^2 - 36y^2$

3) $16x^4 - y^6$

4) $16x^4 - y^4$

5) $8s^3 + 27t^3$

6) $27s^3 + 64t^3$

7) $2x^3 + 64y^3$

8) $8s^3 - 27t^3$

9) $64z^3 - 9t^3$

10) $64y^6 - z^6$

11) $x^2 + 2x + 1$

12) $x^2 + 6x + 8$

13) $x^2 - 2x - 24$

14) $a^4 - 2a^2 - 24$

15) $s^6 - 7s^3 - 8$

16) $s^6 - 26s^3 - 27$

17) $3x^2 + x - 2$

18) $2x^2 + x - 3$

19) $-2x^2 + 8x - 8$

20) $x^2y^2z + 2xy^2z + y^2z$

5.3 Grouping

This method sometimes, but not always, does the job. Here's how it works.

EXAMPLE 1 Factor $10xy + 15y + 4x + 6$.

Solution As written, there are no common factors, but notice that the first and second terms have a common factor of $5y$, while the third and fourth terms have a common factor of 2, giving

$$(10xy + 15y) + (4x + 6) = 5y(2x + 3) + 2(2x + 3).$$

Now these two terms have a common factor of $2x + 3$, which we factor out to get

$$10xy + 15y + 4x + 6 = (2x + 3) \cdot (5y + 2).$$

Notice that the expression is in factored form. ■

EXAMPLE 2 Factor $6ax + 3ay - 4bx - 2by + 10x + 5y$.

Solution You can group in more than one way. Let's try

$$(6ax + 3ay) + (-4bx - 2by) + (10x + 5y)$$
$$= 3a(2x + y) + (-2b)(2x + y) + 5(2x + y)$$
$$= (2x + y) \cdot (3a - 2b + 5). \quad \blacksquare$$

Remarks   a)   In the above example, we could have grouped differently. Noticing that some terms have an x-factor and some a y-factor, we could have done this:

$$6ax + 3ay - 4bx - 2by + 10x + 5y$$
$$= (6ax - 4bx + 10x) + (3ay - 2by + 5y)$$
$$= 2x(3a - 2b + 5) + y(3a - 2b + 5)$$
$$= (3a - 2b + 5)(2x + y)$$

The resulting factorization is the same as before. Always, if grouping is going to work, the resulting factorization will be the same regardless of how you group.

b)   Not all expressions can be factored by grouping; in fact, some expressions cannot be factored **by any method**. If an expression has a prime number of terms, it definitely can't be factored by **grouping**. (Can you explain why?)

c)   The resulting factors may need further factoring afterward.

5.3 Exercises

Factor the following completely:

1) $3ax + 2ay + 3bx + 2by$

2) $2ax + 3ay + 2bx + 3by$

3) $x^4 - x^3y + x - y$

4) $x^5 - x^4y + 2x - 2y$

5) $x^{10} + x^6y^2 + x^4y^3 + y^5$

6) $x^5 - x^3y^2 - x^2y^3 + y^5$

7) $6x^3y - 4xy^3 + 12yx^2 - 8y^3$

8) $3x^3y + 2xy^3 - 6yx^2 - 4y^3$

9) $3x^2 + 5xy + 7x + 3xy + 5y^2 + 7y$

5.4 The Factor Theorem and Long Division

Here it comes, the long-awaited entrance of the ultimate in theoretical sensations. Yes folks, the one, the only, the celebrated Factor Theorem (F.T.). Here is the pearl of wisdom. (Let it rip Frank!)

Factor Theorem

Let $P(x)$ be a polynomial. Let a be any real number. Then $x - a$ is a factor of $P(x)$ if and only if $P(a) = 0$.

Remark　So what's the big deal with this factor theorem? Here's what. Suppose you want to factor a polynomial $P(x)$, and the other methods don't look promising. But suppose you can find some number a such that $P(a) = 0$. (You can do this by trying easy numbers like $a = \pm 1, \pm 2$, etc. to see if you are lucky and find such an a. If you have a graphing calculator, you can see where the graph of the function crosses or touches the x-axis. By zooming in on one such point, you may obtain a number a where $P(a) = 0$.) IF YOU CAN FIND an a with $P(a) = 0$, then you are sure, because of the F.T., that $x - a$ is a factor of $P(x)$. So $P(x) = (x - a) \cdot$ (something). That "something" can be found by long division, and so you've factored $P(x)$!

EXAMPLE 1　Factor $P(x) = x^3 - 2x^2 - 5x + 6$.

Solution　This one looks tough. The earlier methods of factoring would give you nothing. Zip. Zero. Nada. Goose eggs. But notice that if you put $x = 1$, you get $1^3 - 2 \cdot 1^2 - 5 \cdot 1 + 6 = 0$. Aha! So the F.T. says that $x - 1$ is a factor. So

$$P(x) = (x - 1) \cdot \text{(something)}.$$

You're half done. To get the other factor, use long division:

$$x - 1 \overline{)x^3 - 2x^2 - 5x + 6} \qquad \text{(Notice that both expressions are in descending powers!)}$$

Divide the first term into the first term (x into x^3 gives x^2). Put the x^2 on top, multiply it by the **entire divisor** $x - 1$, and subtract it from the dividend $x^3 - 2x^2 - 5x + 6$:

$$
\begin{array}{r}
x^2 \\
x - 1 \overline{)x^3 - 2x^2 - 5x + 6} \\
\underline{x^3 - x^2 } \\
-x^2 - 5x + 6
\end{array}
$$

The easiest way to subtract $x^3 - x^2$ is to **mentally** change the sign of $x^3 - x^2$, getting $-x^3 + x^2$, and **add.** Continue in the same way: x into $-x^2$ gives $-x$, etc., to get:

$$\begin{array}{r}
x^2 - x \\
x - 1 \overline{\smash{\big)}\, x^3 - 2x^2 - 5x + 6} \\
\underline{x^3 - x^2} \\
-x^2 - 5x + 6 \\
\underline{-x^2 + x} \\
- 6x + 6
\end{array}$$

One more time: x into $-6x$ gives -6. So:

$$\begin{array}{r}
x^2 - x - 6 \\
x - 1 \overline{\smash{\big)}\, x^3 - 2x^2 - 5x + 6} \\
\underline{x^3 - x^2} \\
-x^2 - 5x + 6 \\
\underline{-x^2 + x} \\
-6x + 6 \\
\underline{-6x + 6} \\
0
\end{array}$$

We knew that the remainder would be 0 because the F.T. told us so.

Hence $P(x) = (x - 1) \cdot (x^2 - x - 6)$.

So $P(x)$ is factored, but not factored completely as yet because $(x^2 - x - 6)$ can be factored further to give

$$P(x) = (x - 1) \cdot (x - 3) \cdot (x + 2).$$

Success! ∎

EXAMPLE 2 Factor $2x^2 + 3x - 2$.

Solution You could try special formula 5, but it would take too long. Remembering the F.T., all you need is a number a that makes $2a^2 + 3a - 2 = 0$. In ordinary English: you need a solution to $2x^2 + 3x - 2 = 0$. But that's easy, if you recall the quadratic formula,

$$x = \frac{-b \pm \sqrt{b^2 - 4ac}}{2a},$$

where in this context, $a = 2$, $b = 3$, and $c = -2$. Using this formula, you get $x = \dfrac{-3 \pm \sqrt{9 + 16}}{4} = \dfrac{-3 \pm 5}{4} = \dfrac{1}{2}$ or -2. So the F.T. gives us both factors $\left(x - \frac{1}{2}\right)$ and $(x + 2)$, and so

$$2x^2 + 3x - 2 = \left(x - \frac{1}{2}\right) \cdot (x + 2) \cdot (\text{something}).$$

Since the coefficient of the x^2 term on the left is 2, that something **must be 2**, for equality. So

$$2x^2 + 3x - 2 = \left(x - \frac{1}{2}\right)(x + 2)(2) = (2x - 1) \cdot (x + 2). \quad \blacksquare$$

EXAMPLE 3   Can $x^2 + x + 1$ be factored?

Solution The F.T. tells us to check for solutions of the equation $x^2 + x + 1 = 0$. The quadratic formula gives

$$x = \frac{-1 \pm \sqrt{1 - 4}}{2},$$

which is not a real number, because negative numbers don't have square roots among the real numbers. (See Section 1.5 for further explanation.) So there is no real number a to make $a^2 + a + 1$ equal to 0. Hence the F.T. says that $x^2 + x + 1$ **can't be factored**. (Such quadratics are called **irreducible**.) $\blacksquare$

To sum up: You can always use the F.T. to factor quadratics by using the quadratic formula, but for other polynomials $P(x)$, you'll need to be lucky to find a number a that makes $P(a) = 0$. Try $a = \pm 1, \pm 2$, etc.

In real-life problems, scientists and engineers use calculators or computers to find, or approximate, values for a where $P(a) = 0$.

5.4 Exercises

Factor the following expressions, if possible:

1) $x^3 - 3x + 2$

2) $x^3 - 7x + 6$

3) $2x^3 + x + 3$

4) $x^3 - x + 6$

5) $x^2 - 3x - 2$

6) $2x^2 - 3x + 4$

7) $2x^2 - 5x + 4$

8) $24x^2 - 48x - 72$

9) $x^3 - 3x - 2$

10) $x^3 + 2x^2 - 13x + 10$

5.5 Rationalizing Numerators or Denominators Using Conjugates

Consider the expression $\dfrac{x - \sqrt{2}}{5}$. Notice that the numerator has two terms, one of which is a square root. For various reasons, you may not want a square root in the numerator. You can always get rid of it by multiplying and dividing by its conjugate $x + \sqrt{2}$. (You find the **conjugate** by changing the sign between the two terms.) So you will get

$$\frac{x - \sqrt{2}}{5} = \frac{x - \sqrt{2}}{5} \cdot \frac{x + \sqrt{2}}{x + \sqrt{2}} = \frac{\left(x - \sqrt{2}\right) \cdot \left(x + \sqrt{2}\right)}{5 \cdot \left(x + \sqrt{2}\right)}$$

$$= \frac{x^2 - 2}{5 \cdot \left(x + \sqrt{2}\right)}.$$

Remarks

a)   The top is now root-free.

b)   The little devil's conjugate has popped up in the denominator instead. Depending on the problem you're solving, that may not cause any difficulties.

c)   It all boils down to this: you can **exchange** a square root in the top for a square root in the bottom, or vice versa, whichever is better in a particular case.

EXAMPLE 1   Rationalize the denominator of $\dfrac{x^2 - 3}{x + \sqrt{3}}$.

Solution   This means get rid of the root in the bottom. So multiply both top and bottom by the conjugate $x - \sqrt{3}$, giving

$$\frac{x^2 - 3}{x + \sqrt{3}} = \frac{x^2 - 3}{x + \sqrt{3}} \cdot \frac{x - \sqrt{3}}{x - \sqrt{3}} = \frac{\left(x^2 - 3\right)\left(x - \sqrt{3}\right)}{\left(x + \sqrt{3}\right)\left(x - \sqrt{3}\right)}$$

$$= \frac{\left(x^2 - 3\right)\left(x - \sqrt{3}\right)}{\left(x^2 - 3\right)} = \left(x - \sqrt{3}\right). \quad \blacksquare$$

Remark When you're working with a fractional expression, and either the top or bottom has two terms, one (or both) of which is a square root, it is often useful to rationalize it. This comes up in limits, but also in many other cases. If you have such an expression, it **should always pop into your mind that one option is to rationalize**, which may help.

EXAMPLE 2   Rationalize $\dfrac{x^4 - 25}{x - \sqrt{5}}$.

Solution Multiply both top and bottom by the conjugate $x + \sqrt{5}$, giving

$$\frac{x^4 - 25}{x - \sqrt{5}} = \frac{x^4 - 25}{x - \sqrt{5}} \cdot \frac{x + \sqrt{5}}{x + \sqrt{5}} = \frac{(x^4 - 25)(x + \sqrt{5})}{(x^2 - 5)},$$

which can be factored to:

$$= \frac{(x^2 - 5)(x^2 + 5)(x + \sqrt{5})}{(x^2 - 5)} = (x^2 + 5)(x + \sqrt{5}),$$

which is probably easier to deal with. ■

Remark You may have noticed that you can get the same result by factoring the top completely and then canceling:

$$\frac{x^4 - 25}{x - \sqrt{5}} = \frac{(x^2 + 5)(x^2 - 5)}{x - \sqrt{5}} = \frac{(x^2 + 5)(x + \sqrt{5})(x - \sqrt{5})}{(x - \sqrt{5})} = (x^2 + 5)(x + \sqrt{5}).$$

That's true, but you've still got to know how to rationalize (see 6.1 Exercises).

EXAMPLE 3   Rationalize $\dfrac{\sqrt{x + h} - \sqrt{x}}{h}$.

Solution Multiply both top and bottom by the conjugate $\sqrt{x + h} + \sqrt{x}$, giving:

$$\frac{\sqrt{x + h} - \sqrt{x}}{h} = \frac{(\sqrt{x + h} - \sqrt{x})(\sqrt{x + h} + \sqrt{x})}{h(\sqrt{x + h} + \sqrt{x})},$$

which when simplified gives

$$\frac{\sqrt{x + h} - \sqrt{x}}{h} = \frac{x + h - x}{h(\sqrt{x + h} + \sqrt{x})} = \frac{1}{(\sqrt{x + h} + \sqrt{x})}.$$ ■

The following example presents part of the calculation needed to find the derivative of $f(x) = \dfrac{1}{\sqrt{x}}$.

EXAMPLE 4   Write $\dfrac{1}{\sqrt{x+h}} - \dfrac{1}{\sqrt{x}}$ as one fraction, and rationalize the resulting numerator.

Solution   First, use a common denominator to get

$$\frac{\sqrt{x} - \sqrt{x+h}}{\sqrt{x+h}\sqrt{x}}.$$

Now multiply both top and bottom by the conjugate $\sqrt{x} + \sqrt{x+h}$.

$$\frac{\sqrt{x} - \sqrt{x+h}}{\sqrt{x+h}\sqrt{x}} = \frac{\sqrt{x} - \sqrt{x+h}}{\sqrt{x+h}\sqrt{x}} \cdot \frac{\sqrt{x} + \sqrt{x+h}}{\sqrt{x} + \sqrt{x+h}}$$

$$= \frac{x - (x+h)}{\sqrt{x+h}\sqrt{x}(\sqrt{x} + \sqrt{x+h})}$$

$$= \frac{-h}{\sqrt{x+h}\sqrt{x}(\sqrt{x} + \sqrt{x+h})} \quad \blacksquare$$

5.5   Exercises

In Exercises 1–7, rationalize the top or bottom, and simplify.

1) $\dfrac{7}{\sqrt{2} - 1}$

2) $\dfrac{x + \sqrt{2}}{2}$

3) $\dfrac{3}{x - \sqrt{7}}$

4) $\dfrac{x + 1}{x + \sqrt{11}}$

5) $\dfrac{x^2 - 3}{x - \sqrt{3}}$

6) $\dfrac{x^4 - 36}{x + \sqrt{6}}$

7) $\dfrac{x^8 - 9}{x^2 + \sqrt{3}}$

8) Let $f(x) = \dfrac{1}{\sqrt{2x}}$. Calculate $\dfrac{f(x+h) - f(x)}{h}$ and simplify as in Example 4.

5.6 Extracting Factors from Radicals

Radicals can be difficult when computing things, so it usually pays to make them as simple as possible. Extracting factors from under the radical sign is one way of simplifying.

EXAMPLE 1   Simplify $\sqrt[3]{250x^4y^3}$.

Solution Since $(ab)^n = a^n \cdot b^n$ for all n (as long as both sides are defined!), we have $(ab)^{1/3} = a^{1/3} \cdot b^{1/3}$ and $\sqrt[3]{a \cdot b} = \sqrt[3]{a} \cdot \sqrt[3]{b}$. In this example you can write $250x^4y^3 = (125x^3y^3)(2x)$. (We chose the first factor to be a **perfect cube**.) So
$\sqrt[3]{250x^4y^3} = \sqrt[3]{125x^3y^3}\sqrt[3]{2x} = 5xy\sqrt[3]{2x}$. ∎

EXAMPLE 2   Simplify the radical by extracting all that you can from $\sqrt{25x^8y}$.

Solution $\sqrt{25x^8y} = \sqrt{(25x^8)y}$

$$= \sqrt{25x^8}\sqrt{y}$$

$$= 5x^4\sqrt{y}$$ ∎

Remarks
a) Notice that in Example 1 as you "pulled" the factor $125x^3y^3$ out of the radical, it "became" its cube root $5xy$. Similarly, as you "pulled" $25x^8$ out of the radical in Example 2, it "changed into" its square root $5x^4$.

b) When working with square roots and other even-powered roots, you must remember that $\sqrt{a^2} = |a|$, not just a. The next example will illustrate.

EXAMPLE 3   Simplify $\sqrt{36xy^2z^3}$.

Solution $\sqrt{36xy^2z^3} = \sqrt{(36y^2z^2)(xz)}$, where the first factor is a perfect square.

So $\sqrt{36xy^2z^3} = \sqrt{(36y^2z^2)}\sqrt{(xz)}$

$$= \left(\sqrt{36}\sqrt{y^2}\sqrt{z^2}\right)\sqrt{xz}$$

$$= 6|y||z|\sqrt{xz}.$$ ∎

EXAMPLE 4   Simplify $\sqrt[4]{16x^8y^3}$.

Solution $\sqrt[4]{16x^8y^3} = \sqrt[4]{16x^8}\sqrt[4]{y^3}$

$= \sqrt[4]{16}\sqrt[4]{x^8}\sqrt[4]{y^3}$

$= 2|x^2|\sqrt[4]{y^3}$

Here, the stuff in the absolute value sign, x^2, is always nonnegative, and so $|x^2| = x^2$. So the given expression equals

$$2x^2\sqrt[4]{y^3}. \quad \blacksquare$$

EXAMPLE 5   Simplify $\sqrt[5]{32x^{10}y^2}$.

Solution $\sqrt[5]{32x^{10}}\sqrt[5]{y^2} = 2x^2\sqrt[5]{y^2} \quad \blacksquare$

EXAMPLE 6   Simplify $\sqrt{x^2y^6 + 3x^5y^4}$.

Solution The stuff in the radical is not factored yet, so you must do that before you can extract any factors!

$$\sqrt{x^2y^6 + 3x^5y^4} = \sqrt{x^2y^4(y^2 + 3x^3)}$$

$$= \sqrt{x^2y^4}\sqrt{y^2 + 3x^3}$$

$$= |x||y^2|\sqrt{y^2 + 3x^3}$$

$$= |x|y^2\sqrt{y^2 + 3x^3} \quad \blacksquare$$

$\boxed{5.6}$ Exercises

Extract as much as you can from the following roots:

1) $\sqrt{16x^2}$

2) $\sqrt{4x^2 + 8x^4}$

3) $\sqrt[3]{54x^4}$

4) $\sqrt{3x^{12}y}$

5) $\sqrt{5x^4 + 3x^8}$

6) $\sqrt[3]{27x^6y}$

7) $\sqrt{8\pi^2x^3y^4}$

8) $\sqrt[3]{x^7y^4 + x^6y^5}$

9) $\sqrt[4]{x^5y^4 + x^6y^{10}}$

CHAPTER 6

Simplifying Algebraic Expressions

6.1 Working with Difference Quotients

At this point in calculus you're going to be meeting up with expressions called **difference quotients.** Such an expression looks like

$$\frac{f(x + h) - f(x)}{h}.$$

Alternatively, the difference quotient can be written using the variable Δx instead of the variable h. Hence you may be seeing things like

$$\frac{f(x + \Delta x) - f(x)}{\Delta x}.$$

Your ability to work with difference quotients often depends on your ability to expand expressions and cancel common factors. Some of the following examples feature difference quotients, while others are for general practice.

EXAMPLE 1 If $f(x) = x^2$, calculate $\dfrac{f(x + h) - f(x)}{h}$.

Solution Since $f(x + h) = (x + h)^2$,

$$\frac{f(x + h) - f(x)}{h} = \frac{(x + h)^2 - x^2}{h}.$$

Expanding $(x + h)^2$ as $x^2 + 2xh + h^2$ we get

$$\frac{f(x + h) - f(x)}{h} = \frac{(x^2 + 2xh + h^2) - x^2}{h},$$

which simplifies to

$$\frac{f(x+h) - f(x)}{h} = \frac{2xh + h^2}{h} = \frac{h(2x+h)}{h}$$

$$= 2x + h. \quad \blacksquare$$

Remark The last step in Example 1 was to cancel a common factor of h from the numerator and denominator. Notice that we are able to factor and cancel the h since **every** term of the numerator contained a factor of h.

EXAMPLE 2 Let $f(x) = 2x^2 - 3x$. Calculate $\dfrac{f(x + \Delta x) - f(x)}{\Delta x}$.

Solution Since $f(x + \Delta x) = 2(x + \Delta x)^2 - 3(x + \Delta x)$,

$$\frac{f(x + \Delta x) - f(x)}{\Delta x} = \frac{2(x + \Delta x)^2 - 3(x + \Delta x) - (2x^2 - 3x)}{\Delta x}.$$

Expanding $(x + \Delta x)^2$ as $x^2 + 2x\Delta x + (\Delta x)^2$, we get

$$\frac{f(x + \Delta x) - f(x)}{\Delta x} = \frac{2x^2 + 4x\Delta x + 2(\Delta x)^2 - 3x - 3\Delta x - 2x^2 + 3x}{\Delta x},$$

which simplifies to

$$\frac{f(x + \Delta x) - f(x)}{\Delta x} = \frac{4x\Delta x + 2(\Delta x)^2 - 3\Delta x}{\Delta x} = \frac{\Delta x(4x + 2\Delta x - 3)}{\Delta x}$$

$$= 4x + 2\Delta x - 3. \quad \blacksquare$$

Computing difference quotients for functions other than polynomials gets a little tricky. An ability to add and subtract rational expressions becomes essential. Recall from Chapter 1 how to add fractions: $\dfrac{a}{b} + \dfrac{c}{d} = \dfrac{ad + bc}{bd}$. (First form a common denominator, namely bd, then change $\dfrac{a}{b}$ to $\dfrac{ad}{bd}$, and $\dfrac{c}{d}$ to $\dfrac{bc}{bd}$, at which point you can add them to get the result.) The same method applies, no matter how tough these expressions look.

EXAMPLE 3 Simplify $\dfrac{1}{x} - \dfrac{1}{x - 1}$.

Solution $\dfrac{1}{x} - \dfrac{1}{x - 1} = \dfrac{1 \cdot (x - 1) - 1 \cdot x}{x(x - 1)}$

$$= \frac{x - 1 - x}{x(x - 1)} = \frac{-1}{x(x - 1)} \quad \blacksquare$$

EXAMPLE 4   Simplify $\dfrac{3x + y}{x + y} + \dfrac{x - y}{x + 2y}$.

Solution Finding a common denominator and adding gives

$$\frac{(3x + y)(x + 2y) + (x + y)(x - y)}{(x + y)(x + 2y)}.$$

This needs a little "cleaning up." Let's multiply out the top, to get

$$\frac{3x^2 + xy + 6xy + 2y^2 + x^2 + xy - xy - y^2}{(x + y)(x + 2y)}$$

$$= \frac{4x^2 + 7xy + y^2}{(x + y)(x + 2y)}.$$

For most purposes, you would not want to multiply out the denominator. ■

EXAMPLE 5   Simplify $\dfrac{\dfrac{1}{t + 1} - \dfrac{1}{t}}{\dfrac{1}{t + 1} + \dfrac{1}{t}}$.

Solution Working separately with the numerator and denominator,

$$\frac{1}{t + 1} - \frac{1}{t} = \frac{t - (t + 1)}{t(t + 1)} = \frac{-1}{t(t + 1)}$$

$$\frac{1}{t + 1} + \frac{1}{t} = \frac{t + (t + 1)}{t(t + 1)} = \frac{2t + 1}{t(t + 1)}.$$

So

$$\frac{\dfrac{1}{t + 1} - \dfrac{1}{t}}{\dfrac{1}{t + 1} + \dfrac{1}{t}} = \frac{\dfrac{-1}{t(t + 1)}}{\dfrac{2t + 1}{t(t + 1)}} = \frac{-1}{t(t + 1)} \cdot \frac{t(t + 1)}{2t + 1} = \frac{-1}{2t + 1}.$$

This answer is as simple as it gets. ■

EXAMPLE 6 Simplify $\dfrac{x^{-1} + y^{-1}}{(xy)^{-1}}$.

Solution First convert the expression to fractions:

$$\frac{x^{-1} + y^{-1}}{(xy)^{-1}} = \frac{\dfrac{1}{x} + \dfrac{1}{y}}{\dfrac{1}{xy}}$$

$$= \frac{\dfrac{1 \cdot y + 1 \cdot x}{xy}}{\dfrac{1}{xy}} = \frac{\dfrac{y + x}{xy}}{\dfrac{1}{xy}} = \frac{y + x}{1} = y + x. \quad \blacksquare$$

EXAMPLE 7 Simplify $\dfrac{\dfrac{1}{x + h} - \dfrac{1}{x}}{h}$.

Solution $\dfrac{\dfrac{1}{x + h} - \dfrac{1}{x}}{h} = \dfrac{\dfrac{x - (x + h)}{x(x + h)}}{h}$

$$= \frac{\dfrac{x - x - h}{x(x + h)}}{h} = \frac{\dfrac{-h}{x(x + h)}}{h}$$

Notice that both the top and bottom have a factor of h, which can be canceled to obtain

$$\frac{h\left(\dfrac{-1}{x(x + h)}\right)}{h} = \frac{-1}{x(x + h)}. \quad \blacksquare$$

EXAMPLE 8 Simplify $\dfrac{3x^2y^5}{24xy^2}$.

Solution Cancel factors that are common to the numerator and the denominator:

$$\frac{3x^2y^5}{24xy^2} = \frac{xy^3}{8}. \quad \blacksquare$$

EXAMPLE 9 Simplify $\dfrac{5a^2b^3 - 15ab}{100a^2b^2}$.

Solution Remember to avoid "creative canceling." Cancel only those things that are factors of both the **entire top** and the **entire bottom.** Factor the numerator to obtain

$$\frac{5a^2b^3 - 15ab}{100a^2b^2} = \frac{5ab(ab^2 - 3)}{100a^2b^2}.$$

Notice that the numerator and denominator both have a factor of $5ab$. Cancel it to get

$$\frac{ab^2 - 3}{20ab}. \qquad \blacksquare$$

EXAMPLE 10 Simplify $\dfrac{27x^2\left(\dfrac{y}{z}\right)^{-5}}{(3xyz^4)^2}$.

Solution $\dfrac{27x^2\left(\dfrac{y}{z}\right)^{-5}}{(3xyz^4)^2} = \dfrac{27x^2\left(\dfrac{z}{y}\right)^{5}}{9x^2y^2z^8} = \dfrac{27x^2y^{-5}z^5}{9x^2y^2z^8}$

$$= 3y^{-7}z^{-3} = \frac{3}{y^7z^3} \qquad \blacksquare$$

EXAMPLE 11 Let $f(x) = x^2 + \dfrac{1}{x}$.

Calculate: a) $\quad f(x + \Delta x)$

   b) $\quad \dfrac{f(x + \Delta x) - f(x)}{\Delta x}$

Solution a) $\quad f(x + \Delta x) = (x + \Delta x)^2 + \dfrac{1}{x + \Delta x}$

   b) $\quad \dfrac{f(x + \Delta x) - f(x)}{\Delta x} = \dfrac{(x + \Delta x)^2 + \dfrac{1}{x + \Delta x} - \left(x^2 + \dfrac{1}{x}\right)}{\Delta x}$

$$= \frac{x^2 + 2x \cdot \Delta x + (\Delta x)^2 + \dfrac{1}{x + \Delta x} - x^2 - \dfrac{1}{x}}{\Delta x}$$

Cancel the x's, and subtract the two fractions on top. So

$$\frac{f(x + \Delta x) - f(x)}{\Delta x} = \frac{2x \cdot \Delta x + (\Delta x)^2 + \dfrac{x - (x + \Delta x)}{x(x + \Delta x)}}{\Delta x}$$

$$= \frac{2x \cdot \Delta x + (\Delta x)^2 - \dfrac{\Delta x}{x(x + \Delta x)}}{\Delta x}$$

$$= \frac{\Delta x \left(2x + \Delta x - \dfrac{1}{x(x + \Delta x)} \right)}{\Delta x}$$

$$= 2x + \Delta x - \frac{1}{x(x + \Delta x)}. \quad \blacksquare$$

6.1 Exercises

1) Let $f(x) = x^2 + 3x$.

    a) Compute $f(x + h)$.   b) Simplify $\dfrac{f(x + h) - f(x)}{h}$.

2) Let $f(x) = 2x^2 - 2x$.

    a) Compute $f(x + h)$.   b) Simplify $\dfrac{f(x + h) - f(x)}{h}$.

3) Let $g(x) = 2x^3 - x$.

    a) Compute $g(x + h)$.   b) Simplify $\dfrac{g(x + h) - g(x)}{h}$.

4) Let $g(x) = \dfrac{1}{2}x^3$.

    a) Compute $g(x + h)$.   b) Simplify $\dfrac{g(x + h) - g(x)}{h}$.

Simplify the expressions in Exercises 5–13.

5) $\dfrac{1}{x - 1} - \dfrac{1}{x}$ 6) $\dfrac{x}{x - 1} - \dfrac{2}{x}$

7) $\dfrac{x}{x - 1} - \dfrac{x}{x + 1}$ 8) $\dfrac{\dfrac{s + 1}{s - 1} + \dfrac{s - 1}{s + 1}}{\dfrac{1}{s^2 - 1}}$

9) $\dfrac{2(x + h) + 1 - (2x + 1)}{h}$

10) $\dfrac{\dfrac{1}{(x + h)^2} - \dfrac{1}{x^2}}{h}$

11) $x + \dfrac{1}{x + \dfrac{1}{x + \dfrac{1}{x}}}$

12) $\dfrac{\sqrt{2x + 2h} - \sqrt{2x}}{h}$ (*Hint:* Rationalize your way out of this one!)

13) $\dfrac{3a^2b - 27ab^2}{(15a^3b^4)^2}$

14) Let $g(x) = \sqrt{x + 2}$. Calculate $\dfrac{g(x + \Delta x) - g(x)}{\Delta x}$.

15) Let $g(x) = \sqrt{x - 3}$. Calculate $\dfrac{g(x + \Delta x) - g(x)}{\Delta x}$.

16) Calculate the difference quotient for $f(x) = \dfrac{4}{x}$ and simplify.

17) Calculate the difference quotient for $f(x) = \dfrac{1}{\sqrt{x}}$ and simplify.

18) Calculate the difference quotient for $f(x) = \dfrac{\sqrt{2}}{x^2}$ and simplify.

19) Suppose you're on the pike driving from Stockbridge to Boston. Let $d(t)$ be the distance you have traveled at time t. Consider the time interval $(t, t + \Delta t)$. Write an expression for your average speed in that interval.

Cyclic Phenomena: The Six Basic Trigonometric Functions

7

7.1 Angles

The size of an angle can be expressed in several ways. You've used "degrees" for years. They are simple. You've also used "revolutions" when talking about the revolutions per minute (RPM) of an engine. In calculus, "radians" are used as a unit of measure because the rules of calculus are easiest that way. How big is a radian? Here's how big: **it's the angle corresponding to an arc length of 1 in a unit circle.** Look at Figure 7.1. A "unit circle" indicates that the radius $= 1$, and we'll always put the center at $(0, 0)$ for convenience. The angle θ as drawn is 1 radian, because the arc length "subtended" (cut off) by the angle has length $= 1$. By the way, θ is the Greek letter "theta" and is often used for angles.

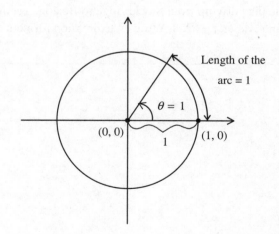

Figure 7.1

Notice that if you double the length of the arc, you double the angle. In fact, in a unit circle, the arc length equals the angle (if you use radians). Since most of you are probably more comfortable using degrees, we must deal with the key question: how do you convert from degrees to radians, and vice versa? Here's the secret:

One revolution = the complete angle at the center in radians, so

= the length of the complete arc

= the circumference of the unit circle

$= 2\pi r$, but $r = 1$, so

$= 2\pi$ (radians, that is)

You recall that there are 360 degrees in one full revolution, so

$$\boxed{360° = 2\pi \text{ radians}}.$$

If you divide both sides of this equation by 2π, you obtain

$$1 \text{ radian} = \frac{360°}{2\pi} \cong 57°,$$

while dividing both sides of the equation by 360 gives

$$1° = \frac{2\pi}{360} \text{ radians} = \frac{\pi}{180} \text{ radians} \cong 0.017 \text{ radians}.$$

We will **always** express angles in **radians**, unless specified differently. Moreover, angles are measured starting at the positive x-axis going counterclockwise if the angle is positive, clockwise if the angle is negative. Typically, we will also leave π simply as π, and will not use any decimal approximation. To a mathematician, leaving π in an answer is fine. In fact it's better than fine because it's exact. Anything else would be less exact. (However, if you go to a hardware store and ask the clerk for π square inches of sheet metal, you will get (and deserve!) some funny stares.)

EXAMPLE 1   Convert 27° to radians.

Solution $360° = 2\pi$ (If things are clear, we usually drop the word "radians.")

$$1° = \frac{2\pi}{360}, \text{ so } 27° = 27 \cdot \frac{2\pi}{360} = \frac{3}{40} \cdot 2\pi = \frac{3\pi}{20}. \quad ■$$

EXAMPLE 2   Convert $-\dfrac{5\pi}{6}$ radians to degrees, and diagram it.

Solution $2\pi = 360°$

$\pi = 180°$

$$-\frac{5}{6} \cdot \pi = -\frac{5}{6} \cdot 180° = -150°.$$

The angle is measured clockwise, because it is negative, and is shown in the figure below.

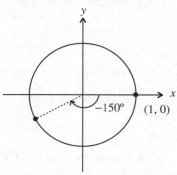

You should commit to memory the following conversion chart since these angles will come up again and again.

Degrees	360	180	90	60	45	30
Radians	2π	π	$\dfrac{\pi}{2}$	$\dfrac{\pi}{3}$	$\dfrac{\pi}{4}$	$\dfrac{\pi}{6}$

7.1 Exercises

1) Change to radian measure:

    a) 120°   b) 270°

    c) 135°   d) 210°

    e) −150°   f) 450°

2) Change the following angles from degree to radian measure:

    a) 75°   b) 225°

    c) 130°   d) 300°

3)   Change to degrees:

  a) $\dfrac{3\pi}{4}$ b) $\dfrac{11\pi}{6}$

  c) $-\dfrac{\pi}{3}$ d) 3π

  e) $\dfrac{9\pi}{2}$ f) $\dfrac{9\pi}{4}$

4)   Change the following angles from radians to degrees:

  a) $\dfrac{5\pi}{6}$ b) $\dfrac{5\pi}{3}$

  c) $-\dfrac{2\pi}{3}$ d) 34π

5)   Draw a large circle and mark each of the angles in Exercise 3.

6)   Draw a large circle and mark each of the angles in Exercise 4.

7.2 Definition of $\sin\theta$ and $\cos\theta$

Consider the unit circle, centered at the origin, with an angle of θ radians, as shown in Figure 7.2.

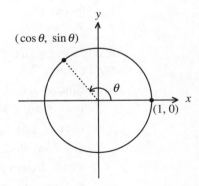

Figure 7.2

(Notice that the angle is measured from the positive x-axis, counterclockwise.) The dotted line defining the terminal side (end) of the angle θ intersects the circle at a point. As the angle θ changes, so do the coordinates of that point, so each of the coordinates is a function of the angle θ. These two functions are very important, and so they have their own names.

DEFINITION

In Figure 7.2 the first coordinate is called **$\cos\theta$** (short for cosine of θ). The second coordinate is called **$\sin\theta$** (short for sine of θ).

Remarks

a) Since this point is on the unit circle, its coordinates must satisfy the equation of that circle: $x^2 + y^2 = 1$, that is, $(\cos\theta)^2 + (\sin\theta)^2 = 1$.

b) To avoid the constant use of parentheses, we write $\cos^n\theta$ to mean $(\cos\theta)^n$; similarly, we write $\sin^n\theta$ to mean $(\sin\theta)^n$. Thus $\cos^2\theta + \sin^2\theta = 1$. We usually write this equation as $\sin^2\theta + \cos^2\theta = 1$.

c) Imagine θ going through values from 0 to 2π. Then the point on the circle goes from $(1, 0)$ counterclockwise around the circle. Thus, its second coordinate, $\sin\theta$, which is the altitude of the point, goes first from 0 up to 1, then down to -1, then back up to 0, as shown in the graph in Figure 7.3.

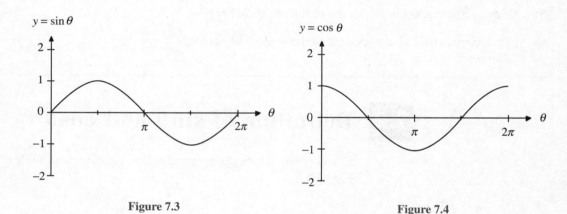

Figure 7.3   Figure 7.4

Meanwhile, the first coordinate, $\cos\theta$, goes from 1 down to -1, and back to 1 as shown in Figure 7.4.

d) Since θ and $\theta + 2\pi$ correspond to the same point on the unit circle, we have $\sin\theta = \sin(\theta + 2\pi)$. This means that the graphs repeat over and over ad infinitum. In fact, $\sin\theta = \sin(\theta + 2k\pi)$ for any whole number k. We say that $\sin\theta$ is **periodic** with period 2π. The same is true of $\cos\theta$. Remember that $2\pi k$ radians, where k is a whole number, is exactly k revolutions.

e) We just happened to use the variable name θ. We could also use x, y, z, w, u, v, ☺, ☹, or any other letter or symbol desired! To be consistent with the function notation $f(x)$ we've come to know and love, we will use the variable x. If we allow x to be any real number, we get the **full** graph of $f(x) = \sin x$ and $g(x) = \cos x$, shown in Figures 7.5 and 7.6.

$y = \sin x$

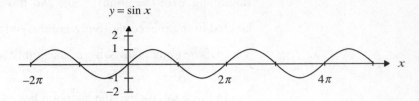

Figure 7.5

$y = \cos x$

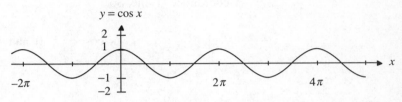

Figure 7.6

f)   You might guess that $\sin x$ and $\cos x$ and other trigonometric functions would be useful for studying waves, daily temperature averages through the seasons, alternating current, electronic signals, and lots of other **cyclic phenomena**. You'd be perfectly right. These functions are the basic tools of the trade. Wave-shaped graphs are said to be **sinusoidal**.

EXAMPLE 1   Evaluate $\cos\dfrac{3\pi}{2}$ and $\sin\dfrac{3\pi}{2}$.

Solution To determine the sine and cosine of some angles is really quite easy. Just remember the diagram below. The angle $\frac{3\pi}{2}$ is 3/4 of the way around the circle, ending up at the point $(0, -1)$. For angles that end up on the coordinate axes, the sine and cosine are given immediately because we know exactly what the coordinates of those points are.

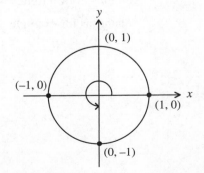

Beginning with the point (1, 0), and traveling counterclockwise, the four points labeled in the preceding figure are the endpoints for the angles 0, $\frac{\pi}{2}$, π, $\frac{3\pi}{2}$, and 2π, whose point is the same as the point corresponding to angle 0.

In this case, for $\frac{3\pi}{2}$ the endpoint has an x-coordinate of 0, and a y-coordinate of -1. Just remember x before y, c before s. The x-coordinate is the **c**osine, and the y-coordinate is **s**ine. So

$$\cos\frac{3\pi}{2} = 0 \text{ and } \sin\frac{3\pi}{2} = -1. \quad \blacksquare$$

EXAMPLE 2   What is $\cos\dfrac{9\pi}{2}$?

Solution First of all, $\frac{9\pi}{2} = 4\pi + \frac{\pi}{2}$, and 4π radians equals two complete revolutions. So the angle $4\pi + \frac{\pi}{2}$ goes twice around the circle (which does nothing), and then another 90°, ending up at the top of the circle, at the point (0, 1). The cosine is the first coordinate of that point. So

$$\cos\frac{9\pi}{2} = 0. \quad \blacksquare$$

Alternative definition of $\sin\theta$ and $\cos\theta$: The definition we just saw for $\sin\theta$ and $\cos\theta$ is called the **unit circle** definition. It is valid for all real numbers θ. However, if $0 < \theta < \frac{\pi}{2}$, there is another, equivalent, definition, called the **right-angled triangle** definition, with which you are probably more familiar. To illustrate, consider any number θ with $0 < \theta < \frac{\pi}{2}$. Notice that θ radians would be an acute angle, so you may draw a right triangle (short for right-angled triangle) with an angle of θ radians, as for example $\triangle ABC$. See Figure 7.7.

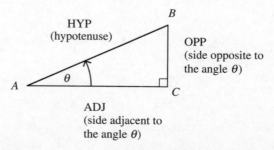

Figure 7.7

Now place the unit circle onto $\triangle ABC$, with its center at A and the positive x-axis along the line segment AC. See Figure 7.8.

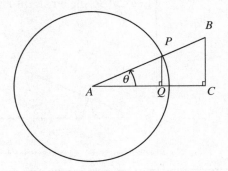

Figure 7.8

Let the intersection of the circle and the line AB be called P. From P drop the perpendicular to AC, to a point Q. Since P is on the unit circle and the terminal line of the angle θ, we have $P = (\cos\theta, \sin\theta)$. Hence, $PQ = \sin\theta$, while $AQ = \cos\theta$ and $AP = 1$. So here we have the two triangles in Figures 7.9(a) and 7.9(b), which are in fact similar.

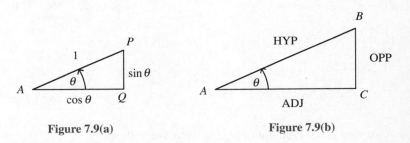

Figure 7.9(a)   **Figure 7.9(b)**

Since they are similar, their sides are in proportion, so

$$\frac{\sin\theta}{1} = \frac{\text{OPP}}{\text{HYP}},$$

and hence

$$\sin\theta = \frac{\text{OPP}}{\text{HYP}}.$$

Similarly

$$\frac{\cos\theta}{1} = \frac{\text{ADJ}}{\text{HYP}},$$

and so

$$\cos\theta = \frac{\text{ADJ}}{\text{HYP}}.$$

DEFINITION

So we have the following alternative definition for $0 < \theta < \frac{\pi}{2}$.

Let $\triangle ABC$ be any right-angled triangle with angle A equal to θ.

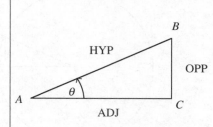

Then $\sin \theta = \dfrac{\text{OPP}}{\text{HYP}} = \dfrac{\text{BC}}{\text{AB}}$,

and $\cos \theta = \dfrac{\text{ADJ}}{\text{HYP}} = \dfrac{\text{AC}}{\text{AB}}$.

7.2 Exercises

1) Evaluate:

    a) $\sin \dfrac{7\pi}{2}$ b) $\cos \dfrac{5\pi}{2}$

    c) $\cos \dfrac{-9\pi}{2}$ d) $\sin 101\pi$

2) Evaluate:

    a) $\cos 12\pi$ b) $\sin \dfrac{5\pi}{2}$

    c) $\sin \dfrac{-9\pi}{2}$ d) $\cos 101\pi$

3) Assuming k is a whole number, evaluate the following:

    a) $\sin(\pi/2 + 2k\pi)$ b) $\cos(-\pi/2 + 2k\pi)$

    c) $\sin k\pi$ d) $\cos k\pi$

4) What is $\cos(\theta + \pi)$ in terms of $\cos \theta$? (*Hint*: Use the unit circle.)

5) Assuming k is a whole number, evaluate the following:

    a) $\cos(\pi/2 + 2k\pi)$ b) $(\sin -\pi/2 + 2k\pi)$

6) a) In the triangle shown, calculate $\sin \theta$ and $\cos \theta$.

    b) Calculate $\sin^2 \theta + \cos^2 \theta$.

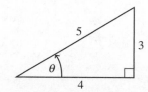

7.3 Special Angles $\left(\dfrac{\pi}{4}, \dfrac{\pi}{6}, \dfrac{\pi}{3}\right)$

EXAMPLE 1 Find $\sin\dfrac{\pi}{4}$.

Solution Since $\dfrac{\pi}{4} = 45°$, the picture looks like

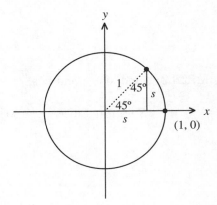

where s is some number.

Hence, the coordinate of the point in question is (s, s), and $s = \sin\frac{\pi}{4} = \cos\frac{\pi}{4}$. Since $\sin^2 x + \cos^2 x = 1$, we have

$$\sin^2\frac{\pi}{4} + \cos^2\frac{\pi}{4} = s^2 + s^2 = 2s^2 = 1.$$

Hence $$s^2 = \frac{1}{2} \text{ and } s = \pm\frac{1}{\sqrt{2}}.$$

Since it is clear that $s > 0$, we have $s = \frac{1}{\sqrt{2}}$, and so:

$$\sin\frac{\pi}{4} = \frac{1}{\sqrt{2}}. \quad \blacksquare$$

Remark **Memorize the Figure 7.10 triangle.** (It will simplify your life.)

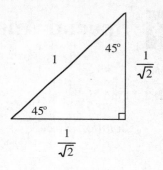

Figure 7.10

EXAMPLE 2 Find $\sin\dfrac{\pi}{6}$ and $\cos\dfrac{\pi}{6}$.

Solution Since $\dfrac{\pi}{6} = 30°$, the picture now looks like

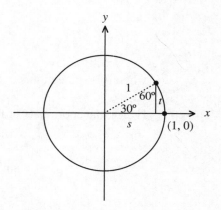

How big is *s*? And *t*? Here's the trick: flip down the triangle along its base, as shown.

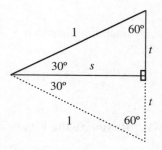

Notice that the result is an equilateral triangle, so the vertical side is also 1. This means $2t = 1$, or $t = \frac{1}{2}$. Now since

$$s^2 + t^2 = 1,$$

$$s^2 + \frac{1}{4} = 1,$$

$$s^2 = \frac{3}{4},$$

and $$s = \pm\frac{\sqrt{3}}{2}.$$

Again, it is clear that $s > 0$, which implies

$$\sin\frac{\pi}{6} = \frac{1}{2} \quad\text{and}\quad \cos\frac{\pi}{6} = \frac{\sqrt{3}}{2}. \quad\blacksquare$$

Remark **Memorize the Figure 7.11 triangle also.** (This is the last one—we promise!)

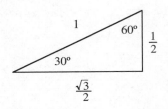

Figure 7.11

The two triangles we asked you to memorize can be used to determine trigonometric values for many more angles. Consider the following example.

EXAMPLE 3 Find $\sin\dfrac{-2\pi}{3}$.

Solution Since $-\frac{2\pi}{3} = -120°$, the triangle in the figure below is exactly the 30°, 60°, 90° triangle shown in Figure 7.11.

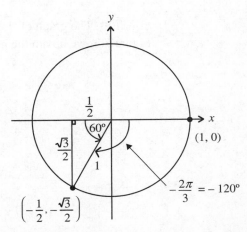

Hence, $\sin\frac{-2\pi}{3} = -\frac{\sqrt{3}}{2}$. (See the diagram above.) $\quad\blacksquare$

7.3 Exercises

Evaluate the following:

1) $\sin\dfrac{3\pi}{4}$

2) $\sin\dfrac{7\pi}{4}$

3) $\cos\dfrac{-\pi}{6}$

4) $\cos\dfrac{-3\pi}{4}$

5) $\cos\dfrac{-11\pi}{4}$

6) $\sin\dfrac{-5\pi}{4}$

7) $\sin\dfrac{2\pi}{3}$

8) $\cos\dfrac{13\pi}{6}$

9) $\sin\dfrac{4\pi}{3}$

10) $\cos\dfrac{-5\pi}{3}$

11) $\cos\dfrac{7\pi}{6}$

12) $\sin\dfrac{29\pi}{6}$

7.4 Graphs Involving sin *x* and cos *x*

Remember Chapter 4, where we shifted graphs up and down, left and right, and so on? We can of course do that with sin *x* and cos *x* also.

EXAMPLE 1   On $[0, 2\pi]$, graph $\sin x$, $1 + \sin x$, and $\sin\left(x - \dfrac{\pi}{2}\right)$.

Solution The graph of $1 + \sin x$ is that of sin *x* shifted up a distance of 1. Recalling Figure 7.3, we obtain the graph:

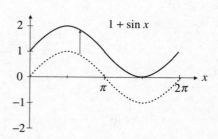

The graph of $\sin\left(x - \frac{\pi}{2}\right)$ is that of sin x shifted right a distance of $\frac{\pi}{2}$.

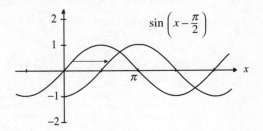

A shift to the left or right of a wave-shaped graph, such as sin x or cos x, is called a **phase shift**.

Now for vertical scaling, or stretching and compressing, consider the graph of $y = \sin x$ on the interval $[0, 2\pi]$ shown in Figure 7.12. How do you stretch it away from the x-axis so it is twice as high above the axis and twice as low below the axis? Simple: y needs to be twice as big, so $y = 2\sin x$ will do it.

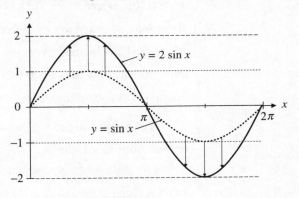

Figure 7.12

To compress the graph so that it lies between $y = -\frac{1}{2}$ and $y = \frac{1}{2}$, simply multiply the function sin x by $\frac{1}{2}$. Figure 7.13 illustrates $y = \frac{1}{2}\sin x$.

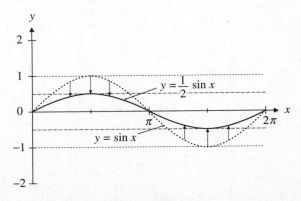

Figure 7.13

In general, when given the graph of $f(x)$, the graph of $k \cdot f(x)$ is that of $f(x)$ stretched if $k > 1$ and compressed if $0 < k < 1$. What happens if $k < 0$? Well, let's see. The graph of $y = -\sin x$ is shown in Figure 7.14.

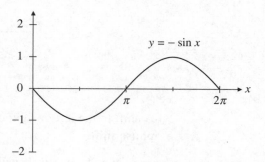

Figure 7.14

It's the mirror reflection of the graph of $y = \sin x$ about the x-axis. In general, for $k < 0$, the graph of $k \cdot f(x)$ is obtained from the graph of $f(x)$ through reflection about the x-axis and a scaling depending on the size of $|k|$. For example, the graph of $-2f(x)$ is obtained by taking the graph of $f(x)$, stretching by a factor of 2, and then reflecting about the x-axis. The graph of $-\frac{1}{2}f(x)$ is obtained by taking the graph of $f(x)$, shrinking it to $\frac{1}{2}$ its vertical size, and then reflecting about the x-axis.

The **amplitude** of a sinusoidal graph is equal to $\frac{1}{2}$ of the distance from the top to the bottom of the waves. For example, $y = 2 \sin x$ has amplitude 2, and $y = 1 - 3 \sin x$ has an amplitude of 3.

Now let's consider horizontal stretching and compressing. When graphing $\sin x$, as x goes from 0 to 2π, $\sin x$ goes from 0 to 1, to -1, back up to 0, as shown in Figure 7.15.

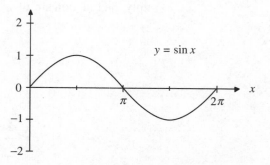

Figure 7.15

(sin x is said to go through **one cycle** on $[0, 2\pi]$, because the rest is repetition.) Old news! Ho hum! Now, what happens to sin $2x$? Well, as x goes from 0 to 2π, $2x$ goes from 0 to 4π, and the sine of it (i.e., sin $2x$) will have the chance to go through two cycles, as in Figure 7.16.

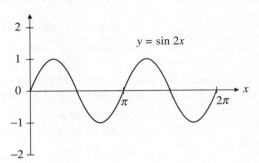

Figure 7.16

On the other hand, as Figure 7.17 shows, for $\sin\frac{1}{2}x$ as x goes from 0 to 2π, $\frac{1}{2}x$ goes from 0 to π (i.e., through only half a cycle).

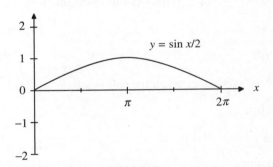

Figure 7.17

In general, here's how you go from $y = \sin x$ to $y = \sin ax$. (In fact, the same method holds in general, for going from $y = f(x)$ to $y = f(ax)$.)

1) If $a > 1$, you compress the graph toward the y-axis. If a is a positive integer, then the graph of $y = \sin ax$ has a complete **oscillations**, or cycles, in the interval $[0, 2\pi]$. (This is also true for $y = \cos ax$.)

2) If $0 < a < 1$, you stretch the graph away from the y-axis. The graph of $y = \sin ax$ has a complete oscillations, or cycles, in the interval $[0, 2\pi]$. For example, if $a = \frac{1}{5}$, we would have one-fifth of the total cycle.

3) If $a < 0$, the graph is reflected about the y-axis, and then compressed or stretched by a factor of $|a|$.

Don't forget:

1)   You can't "factor out" the 2 and say that $\sin(2x)$ and $2 \sin(x)$ are the same thing even if the devil tempts you to. (How simple life would be!) Test it with $x = \frac{\pi}{2}$.

2)   You also can't say that $\sin(x + y)$ and $\sin x + \sin y$ are equal, except by lying or fooling yourself.

Remark   We saw that sinusoidal graphs are made of cycles that just keep repeating. The length of such a cycle is called the **period** of the function, and the reciprocal of the period is called the **frequency**. So $\sin x$ has a period of 2π and a frequency of $\frac{1}{2\pi}$, while $\sin 3x$ has period $\frac{2\pi}{3}$ and a frequency of $\frac{3}{2\pi}$.

7.4 Exercises

1)   From the graph of $y = \sin x$, graph the following on $(0, 2\pi)$:

   a) $y = 3 \sin x$ b) $y = -\frac{1}{2} \sin x$

   c) $y = -2 \sin x$

2)   From the graph of $y = \sin x$, graph the following on $(0, 2\pi)$:

   a) $y = 2.03 \sin x$ b) $y = -\frac{1}{4} \sin x$

   c) $y = -4 \sin x$

3)   Graph the following functions over the interval $(0, 2\pi)$:

   a) $y = \sin 4x$ b) $y = \cos 3x$

   c) $y = \sin \frac{1}{4} x$

4)   Graph the following functions over the interval $(0, 2\pi)$:

   a) $y = \sin 2x$ b) $y = \cos \frac{3}{2} x$

   c) $y = \sin \pi x$

5)   Graph all of the following functions on $(0, 2\pi)$ on the same set of axes.

   a) $y = 2 \sin x$ b) $y = 2 \sin(x - \pi)$

   c) $y = 2 \sin(x - \pi) + 5$

6) Graph all of the following functions on $(0, 2\pi)$, on the same set of axes.

   a) $y = 2\cos x$ b) $y = 2\cos 2x$

   c) $y = 2\cos 2x + 2$

7) On $(0, 4\pi)$ graph $y = 1 - 2\cos x$.

8) On $(-2\pi, 4\pi)$ graph $y = 1 - 3\sin x$.

9) Graph $y = 1 + 2\cos 3x$, and find its amplitude, period, and frequency.

10) Graph $y = 3\sin(2x) - 1$, and find its amplitude, period, and frequency.

11) Graph $y = \sin 2\pi x$, and find its period and frequency.

12) Graph $y = 5\sin(3x) + 2$, and find its period and frequency.

13) Graph $\sin x$ on $(0, 2\pi)$ and $\sin(-x)$ on $(-2\pi, 0)$.

14) The graph of $f(x) = A\sin x$ oscillates between A and $-A$, taking on those particular values when $x = \frac{\pi}{2} + 2k\pi$ and $x = \frac{3\pi}{2} + 2k\pi$ respectively. What if A is not constant but a function of x? What do you think the graph of $g(x) = A(x)\sin x$ does? Graph $\sqrt{x}$ and $\sqrt{x}\sin x$.

7.5 The Other Trigonometric Functions

Having defined the functions $f(x) = \sin x$ and $f(x) = \cos x$ for any real number x, there are four other basic trigonometric functions that you will encounter on your travels:

1) $\tan x = \dfrac{\sin x}{\cos x}$ (called **tangent**)

2) $\cot x = \dfrac{\cos x}{\sin x}$ (called **cotangent**)

3) $\sec x = \dfrac{1}{\cos x}$ (called **secant**)

4) $\csc x = \dfrac{1}{\sin x}$ (called **cosecant**)

Remark At certain values of x, the denominator is zero; hence at those values of x, that function is not defined. For example, $\tan x$ is not defined at $\frac{\pi}{2}$, because $\cos\frac{\pi}{2} = 0$.

7.5 Exercises

1) Evaluate, if they are defined:

   a) $\tan\dfrac{5\pi}{2}$

   b) $\csc\dfrac{5\pi}{2}$

   c) $\sec\dfrac{5\pi}{2}$

   d) $\cot\dfrac{5\pi}{2}$

   e) $\tan\dfrac{-\pi}{2}$

   f) $\sin\dfrac{2\pi}{3}$

   g) $\sec\dfrac{-\pi}{2}$

   h) $\cot\dfrac{-\pi}{2}$

2) Evaluate:

   a) $\csc\dfrac{3\pi}{4}$

   b) $\tan\dfrac{5\pi}{6}$

   c) $\cot\dfrac{-2\pi}{3}$

   d) $\sec\dfrac{7\pi}{4}$

   e) $\cot\dfrac{-\pi}{3}$

   f) $\sec\dfrac{-3\pi}{4}$

   g) $\tan\dfrac{5\pi}{4}$

   h) $\csc\dfrac{2\pi}{3}$

3) Evaluate:

   a) $\tan\dfrac{3\pi}{4}$

   b) $\csc\dfrac{5\pi}{6}$

   c) $\sec\dfrac{-2\pi}{3}$

   d) $\cot\dfrac{7\pi}{4}$

   e) $\tan\dfrac{-\pi}{3}$

   f) $\csc\dfrac{-3\pi}{4}$

   g) $\sec\dfrac{5\pi}{4}$

   h) $\cot\dfrac{2\pi}{3}$

4)   Notice that $\dfrac{1}{x}$ gets large if x gets small—for example, $\dfrac{1}{\frac{1}{1000}} = 1000$. Also recall that $\csc x = \dfrac{1}{\sin x}$. Thus if we draw the graph of $\sin x$ from 0 to $\dfrac{\pi}{2}$, we can envision the graph of $\csc x$ as shown below. (When $\sin x$ is small, $\csc x$ is large.

As $\sin x$ gets close to 0, $\csc x$ increases without bound.)

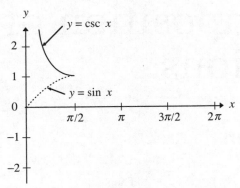

Complete this sketch for x in $[0, 2\pi]$—i.e., graph $\csc x$ on $[0, 2\pi]$ (except at certain points where it is not defined).

5)   Similarly graph $\sec x$ on $[0, 2\pi]$, and then for several periods (except at certain ...).

6)   Graph $\tan x$ on $[-\pi, \pi]$ (except at certain ...).

7)   Graph $\cot x$ on $[0, 2\pi]$ (except at certain ...).

8)   Using the unit circle, estimate $\cos(-1.2)$. (A rough approximation is fine.)

(*Hint:* $\dfrac{\pi}{2} \approx \dfrac{3}{2} = 1.5$, so 1.2 is an angle a little smaller than 90°.)

9)   Estimate $\sin(3.0)$.

10)

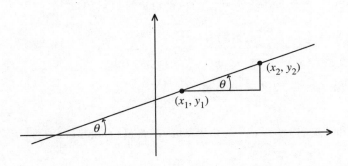

Consider the line L with equation $y = mx + b$ and angle of inclination θ as shown.

a)   Calculate $\tan\theta$ in terms of the coordinates shown.

b)   Calculate the slope m of the line in terms of those coordinates.

c)   Notice anything?

CHAPTER

Composition and Decomposition of Functions

8.1 Composition

Consider the function $f(x) = \sqrt{x + 1}$. Then

$$f(0) = 1,$$
$$f(1) = \sqrt{2},$$
$$f(x^4) = \sqrt{x^4 + 1},$$
$$f(\sin x) = \sqrt{\sin x + 1},$$
$$f(g(x)) = \sqrt{g(x) + 1},$$

and
$$f(\text{whatever}) = \sqrt{\text{whatever} + 1}.$$

Get it? Got it? Good!

EXAMPLE 1   Let $f(x) = x^2 + 2x + \pi$ and $g(x) = x^3$. Then what is $f(g(x))$?

Solution $f(g(x)) = (g(x))^2 + 2g(x) + \pi$
$$= (x^3)^2 + 2x^3 + \pi$$
$$= x^6 + 2x^3 + \pi. \quad \blacksquare$$

EXAMPLE 2   If $f(x) = \tan x$ and $g(x) = \sqrt{x + 1}$, then what is $f(g(x))$ and $g(f(x))$?

Solution a) $f(g(x)) = \tan(g(x)) = \tan \sqrt{x + 1}.$
    b) $g(f(x)) = \sqrt{f(x) + 1} = \sqrt{\tan x + 1}.$

Notice that $f(g(x)) \neq g(f(x))$. $\blacksquare$

DEFINITION | $f(g(x))$ is called the **composition of f with g**. Its symbol is $f \circ g$, and its evaluation at a point x is denoted $(f \circ g)(x)$. So $(f \circ g)(x)$ is another way of writing $f(g(x))$.

EXAMPLE 3 Let $f(x) = \dfrac{1}{x}$, and $g(x) = \sin x$. Find $(f \circ g)(x)$.

Solution $(f \circ g)(x) = f(g(x)) = \dfrac{1}{g(x)} = \dfrac{1}{\sin x} = \csc x.$ ∎

EXAMPLE 4 Let $f(x) = \dfrac{1}{x + 2}$, and $g(x) = x^2 - 1$. Find $(f \circ g)(x)$ and $(g \circ f)(x)$.

Solution a) $(f \circ g)(x) = f(g(x)) = \dfrac{1}{g(x) + 2} = \dfrac{1}{x^2 - 1 + 2} = \dfrac{1}{x^2 + 1}.$

b) $(g \circ f)(x) = (f(x))^2 - 1 = \left(\dfrac{1}{x + 2}\right)^2 - 1.$

Also in this example, $f \circ g \neq g \circ f.$ ∎

We can define $f \circ g \circ h$ in a similar fashion, and also $f \circ g \circ h \circ k$, and so on, and so on.

EXAMPLE 5 Let $f(x) = x^2$, $g(x) = \sin x$, and $h(x) = 2x + 1$. What is the composition $(f \circ g \circ h)(x)$?

Solution $(f \circ g \circ h)(x) = f(g(h(x)))$

$$= f(g(2x + 1)) = f(\sin(2x + 1))$$

$$= (\sin(2x + 1))^2 = \sin^2(2x + 1).$$ ∎

EXAMPLE 6 Let $f(x) = x^3$, $g(x) = \cos x$, $h(x) = \sqrt{x}$, and $k(x) = x + 2$.
Find $(f \circ g \circ h \circ k)(x)$.

Solution $(f \circ g \circ h \circ k)(x) = f(g(h(k(x))))$

$$= f(g(h(x + 2))) = f(g(\sqrt{x + 2}))$$

$$= f(\cos \sqrt{x + 2})$$

$$= (\cos \sqrt{x + 2})^3 = \cos^3 \sqrt{x + 2}.$$ ∎

EXAMPLE 7 Let $f(x) = \cos x$ and $g(t) = t^2$. Find the function $(f \circ g)(t)$.

Solution Notice here that g is a function of t, not x, but that's no problem. You simply substitute $g(t)$ into $f(x)$ to get:

$$(f \circ g)(t) = f(g(t)) = f(t^2) = \cos(t^2) = \cos t^2.$$ ∎

8.1 Exercises

1) Given the functions:

$f(x) = x^3$, $g(x) = \cos x$, $s(t) = 2t + 1$, $h(x) = \sin x - 4x$, find the following composition functions:

a) $(f \circ g)(x)$

b) $(f \circ s)(t)$

c) $(f \circ h)(x)$

d) $(g \circ f)(x)$

e) $(g \circ g)(x)$

f) $(g \circ h)(x)$

2) Given the functions:

$f(x) = x^2 + 1$, $g(x) = \sin x$, $s(t) = 2t - 3$, find the following composition functions:

a) $(f \circ g)(x)$

b) $(f \circ s)(t)$

c) $(g \circ s)(t)$

d) $(g \circ f)(x)$

e) $(g \circ g)(x)$

3) Suppose that $f(x) = x^2 - 2x$, $g(x) = \sqrt{x}$, and $h(x) = \tan x$. Find:

a) $(f \circ g \circ h)(x)$

b) $(f \circ h \circ g)(x)$

c) $(g \circ h \circ f)(x)$

d) $(g \circ f \circ h)(x)$

e) $(f \circ f \circ g)(x)$

f) $(g \circ g \circ f)(x)$

4) Suppose that $f(x) = x^3 + 4x$, $g(x) = \sqrt{x + 1}$, and $h(x) = \cos x$. Find:

a) $(f \circ g \circ h)(x)$

b) $(f \circ h \circ g)(x)$

8.2 Decomposition

When applying the chain rule to a function, it is necessary to decompose that function, that is, to write it as a composition of simpler functions. In other words, a function that is a **composite** of two or more other functions will have to be recognized as such. By the way, in the function $f(g(x))$, f is called the **outer function,** and g is called the **inner function.** Taking derivatives of functions that are composites requires using the chain rule. It is important to let f be the **outermost** function (because there may be several ways of decomposing). How can you find the outermost function? Answer: if you were to evaluate the function at some point, the last operation you would do corresponds to the outermost function.

EXAMPLE 1 Decompose the function $y(x) = (x^2 + 1)^5$.

Solution If you were to evaluate this function (at $x = 0$ for example), the last operation would be to take the fifth (in a manner of speaking). So $f(x) = x^5$ is the outermost function. Clearly $g(x) = x^2 + 1$ is the inner function and $(f \circ g)(x) = (g(x))^5 = (x^2 + 1)^5$. Hence $f(x) = x^5$, $g(x) = x^2 + 1$ gives the desired decomposition. ∎

EXAMPLE 2 Decompose the function $y(x) = \cos^3 x$.

Solution Since $\cos^3 x$ means $(\cos x)^3$, the last operation would be to take the cube. So $f(x) = x^3$ is the outer function and $g(x) = \cos x$ is the inner function. ∎

EXAMPLE 3 Decompose the function $y(x) = \tan \sqrt{x}$.

Solution Here you would take x, take its root, and then take tan of the result. So $f(x) = \tan x$ is the outer function, and $g(x) = \sqrt{x}$ is the inner function. ∎

EXAMPLE 4 Decompose the function $y(x) = \cos \sqrt{x^2 + 1}$.

Solution The outermost function is $f(x) = \cos x$, and the inner function is $g(x) = \sqrt{x^2 + 1}$. (By the way, for the next step in using the chain rule, the function $g(x) = \sqrt{x^2 + 1}$ must itself be decomposed, which yields the outer function $\sqrt{x}$ and the inner function $x^2 + 1$.) ∎

8.2 Exercises

Find decompositions for the following functions (i.e., find f and g such that $y = f \circ g$, and f is the outermost function):

1) $y(x) = \tan^2 x$

2) $y(x) = \csc^3 x$

3) $y(x) = (x^3 - 1)^2$

4) $y(x) = (x^5 + 3)^2$

5) $y(x) = \cos x^5$

6) $y(x) = \sin x^3$

7) $y(x) = \sin \sqrt{x}$

8) $y(x) = \tan \dfrac{1}{x}$

9) $y(x) = \left(\sqrt[5]{x} - 1\right)^{2/3}$

10) $y(x) = \sin \sqrt{x + 1}$

11) $y(x) = \tan^3 2x$

12) $y(x) = \cos(x^3 - 2)^{2/7}$

Equations of Degree 1 Revisited

9.1 Solving Linear Equations Involving Derivatives

You're asking: why are we doing this again? Well, in calculus, when you are using the method of implicit differentiation, what sometimes results is a rather nasty looking equation that contains lots of x's and y's and the derivative $\dfrac{dy}{dx}$ (or y', depending on your notation). However, if you have carried out the differentiation correctly, you will notice that $\dfrac{dy}{dx}$ appears only to the first power. That is, there are no terms that contain $\left(\dfrac{dy}{dx}\right)^2$, or $\left(\dfrac{dy}{dx}\right)^3$, or even $\sqrt{\dfrac{dy}{dx}}$. This means that your equation is really a linear equation in the variable $\dfrac{dy}{dx}$. All the methods shown in Section 3.1 can be used here. Just remember that $\dfrac{dy}{dx}$ (or y') is just a symbol. It's a variable like $x, y, z, t,$ or w. Treat it like one.

EXAMPLE 1 Solve for $\dfrac{dy}{dx}$: $2 + 4\dfrac{dy}{dx} = \dfrac{dy}{dx} + 1.$

Solution Isolate the variable we wish to solve for, which in this case is the derivative $\dfrac{dy}{dx}$. Bring all the terms with $\dfrac{dy}{dx}$ to the left by subtracting $\dfrac{dy}{dx}$ from both sides to give

$$2 + 3\dfrac{dy}{dx} = 1.$$

Now subtract 2 from each side of the equation to give

$$3\dfrac{dy}{dx} = -1.$$

Lastly we divide by 3 to get the solution:

$$\dfrac{dy}{dx} = -\dfrac{1}{3}. \quad \blacksquare$$

EXAMPLE 2   Solve for y': $2x + 3y' = 3x - 5y'$.

Solution Isolate the variable we wish to solve for, which in this case is the derivative y', by bringing all the terms with y' to the left, and all the others to the right. Do this by first adding $5y'$ to both sides of the equation. This gives us

$$2x + 8y' = 3x.$$

Now subtract $2x$ from both sides of the equation to give

$$8y' = x.$$

Finally, by dividing by 8, we have isolated the desired variable y':

$$y' = \frac{x}{8}. \quad \blacksquare$$

EXAMPLE 3   Solve for $\dfrac{dy}{dx}$: $x + 2y\dfrac{dy}{dx} = -\dfrac{dy}{dx} + y$.

Solution Again, isolate the variable we wish to solve for, which in this case is the derivative $\dfrac{dy}{dx}$. Bring all the terms with $\dfrac{dy}{dx}$ to the left, and all others to the right. First, we add $\dfrac{dy}{dx}$ to both sides to give

$$x + 2y\frac{dy}{dx} + \frac{dy}{dx} = y.$$

Now subtract x from each side of the equation to give

$$2y\frac{dy}{dx} + \frac{dy}{dx} = y - x.$$

Now we factor out the $\dfrac{dy}{dx}$ to give us

$$\frac{dy}{dx}(2y + 1) = y - x,$$

and dividing by $(2y + 1)$ gives the desired result:

$$\frac{dy}{dx} = \frac{y - x}{2y + 1}. \quad \blacksquare$$

Things can get a bit more complicated when the equation contains other variables such as x or y. The method of solution, however, is exactly the same. Just be very careful when carrying out the steps. Write neatly and slowly.

EXAMPLE 4   Solve for $\dfrac{dy}{dx}$: $5xy + 4\dfrac{dy}{dx} = 3x^2 - 2xy^2\dfrac{dy}{dx}$.

Solution Bring all the terms with $\dfrac{dy}{dx}$ to the left by adding $2xy^2\dfrac{dy}{dx}$ to both sides of the equation. We get

$$5xy + 4\frac{dy}{dx} + 2xy^2\frac{dy}{dx} = 3x^2.$$

Now bring all the other terms to the right by subtracting $5xy$ from both sides to get

$$4\frac{dy}{dx} + 2xy^2\frac{dy}{dx} = 3x^2 - 5xy.$$

On the left, take out the common factor $\dfrac{dy}{dx}$, to get

$$\frac{dy}{dx}(4 + 2xy^2) = 3x^2 - 5xy.$$

Division by $(4 + 2xy^2)$ gives the result:

$$\frac{dy}{dx} = \frac{3x^2 - 5xy}{4 + 2xy^2}. \quad \blacksquare$$

Remark Notice that when you add or subtract to move terms to one side or the other, they change signs as they "cross the equal sign." Also notice that when you divide (e.g., by $4 + 2xy^2$ in the last example), you **do not** change the sign.

EXAMPLE 5   Solve for y': $x^2y' - 2xy + 2xyy' = (x^2 + 1)y'$.

Solution Again, move all the terms with y' to the left, and all others to the right:

First, subtract $(x^2 + 1)y'$ from both sides, to get
$$x^2y' - 2xy + 2xyy' - (x^2 + 1)y' = 0.$$

Then add $2xy$ to both sides of the equation to give
$$x^2y' + 2xyy' - (x^2 + 1)y' = 2xy.$$

Now, factor out the common factor y':
$$(x^2 + 2xy - x^2 - 1)y' = 2xy.$$

Notice that the x^2-terms cancel, so divide by what's left, $(2xy - 1)$, to get:

$$y' = \frac{2xy}{(2xy - 1)}. \quad \blacksquare$$

Note   You can't cancel the *2xy* term in any way! (If you don't know why, look up "creative canceling" in the Index.)

9.1 Exercises

1)   Solve for $\dfrac{dy}{dx}$: $4 + 6\dfrac{dy}{dx} = 2 + 4\dfrac{dy}{dx}$.

2)   Solve for y': $-y' - 1 = 2y' - 6$.

3)   Solve for $\dfrac{dy}{dx}$: $x + y\dfrac{dy}{dx} = 2x + 4y\dfrac{dy}{dx}$.

4)   Solve for y': $x^2 - x^2y' - 1 = xyy' - 6x$.

5)   Solve for $\dfrac{dy}{dx}$: $2xy + x^2\dfrac{dy}{dx} + 3x^2y^3 + 3x^3y^2\dfrac{dy}{dx} = 0$.

6)   Solve for $\dfrac{dy}{dx}$: $xy + x\dfrac{dy}{dx} + 2xy^2 + 2x^2y\dfrac{dy}{dx} = 3x - 2y\dfrac{dy}{dx}$.

7)   Solve for $\dfrac{dy}{dx}$: $y^2\dfrac{dy}{dx} - 2xy = x + 4x^2\dfrac{dy}{dx}$.

Word Problems, Algebraic and Trigonometric

10.1 Algebraic Word Problems

The methods of finding maxima and minima are some of the most useful mathematical tools in calculus. After all, in many situations, there is a desire to minimize cost, maximize gain, maximize volume for a given amount of material, or minimize the material necessary to enclose a given volume. These methods are also called **optimization** methods, because they address one of the main goals in engineering, science, and business: to do something in the best possible way.

The so-called max-min problems are phrased in a way that is intended to reflect a real-life situation as it may occur to an engineer, for example. First, it is necessary to interpret the situation, to see what quantities are involved, and what questions need to be asked and answered. You won't be told to optimize this specific function, but rather you'll first have to figure out, on your own, what function needs to be optimized. The point is that you need to first understand the problem as given, and to figure out what the variables are and how they are related to each other as functions. "A picture is worth a thousand words," and nine times out of ten, the situation is made easier to understand with a well-drawn diagram. You should make a habit of interpreting word problems through visualization. What does the situation look like in your mind's eye? Now draw it.

EXAMPLE 1 A rectangular field is to be fenced off next to a straight river, with fencing on three sides, with the river edge making the fourth side. Exactly 100 ft of fencing is to be used. Express the area of the field as a function of its width.

Solution First draw a diagram and label the edges.

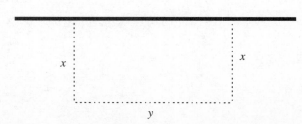

Let x = the width of the field.

Let y = the length of the field.

Note that $x + y + x = 2x + y = 100$, so $y = 100 - 2x$.

Hence the area $A = xy = x(100 - 2x) = 100x - 2x^2$. ∎

EXAMPLE 2 A swimming pool is in the shape of a square with a semicircle at each of two opposite edges. Express the perimeter and area of the pool as a function of the diameter of the semicircles.

Solution

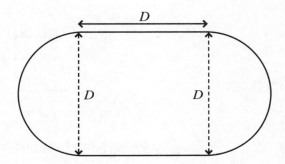

a) The circumference of a circle is $2\pi r$, so the sum of the arc lengths of the two semicircles is $2\pi r$, which equals πD. The square contributes $2D$, so the perimeter of the pool is $\pi D + 2D$.

b) The area of a circle is πr^2, so the sum of the area of the two semicircles is $\pi r^2 = \dfrac{\pi D^2}{4}$, making the area of the pool equal to $\dfrac{\pi D^2}{4} + D^2$. ∎

EXAMPLE 3 A cylindrical tin can has height h cm and radius r cm. Its volume is 32 cm³. Express h as a function of r, and vice versa.

Solution Again, first draw a picture and label the variables.

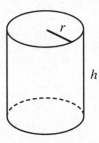

$$V = (\text{area of end})(\text{height}) = (\pi r^2)h = 32.$$

$$\therefore h = \frac{32}{\pi r^2} \text{cm.}$$

Also $\pi r^2 = \dfrac{32}{h}$, so $r^2 = \dfrac{32}{\pi h}$, and hence

$$r = \sqrt{\dfrac{32}{\pi h}}\,\text{cm.} \quad \blacksquare$$

EXAMPLE 4   A 10-in. wire is cut into two pieces. One of the pieces, of length x, is bent into a square. The other piece is bent into a circle. What is the total area of the two shapes as a function of x?

Solution Since one piece is of length x in., the other piece is of length $10 - x$ in.

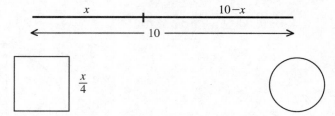

The piece of length x is bent into a square of side $\dfrac{x}{4}$, and therefore the area of the

square is $\dfrac{x^2}{16}$ in^2. The second piece of wire is of length $10 - x$, which forms the cir-

cumference of the circle. Hence we have $10 - x = 2\pi r$, and $r = \dfrac{10 - x}{2\pi}$. So the area

of the circle is $\pi r^2 = \pi\Big(\dfrac{10 - x}{2\pi}\Big)^2 = \dfrac{(10 - x)^2}{4\pi}$. The sum of the areas is

$$A = \dfrac{x^2}{16} + \dfrac{(10 - x)^2}{4\pi}\,\text{in}^2. \quad \blacksquare$$

Recall that when we say that **the variable y is proportional to the variable x**, we mean that $y = kx$ for all x and y and for some fixed number k, which is called the **proportionality constant**. Instead of saying "is proportional to" we can also say "varies as." We say that the variable y is **inversely proportional** to the variable x if $y = \dfrac{k}{x}$ for all x and y, and for some constant k. When y is said to follow the **inverse square law** with respect to x, we mean that $y = \dfrac{k}{x^2}$ for some constant k. There are many situations in real life where quantities obey these relationships. Here are two examples.

EXAMPLE 5   The cost C of building a highway through a certain section of the country is proportional to its length L. A 2.5-mile section costs $1 million. Express the cost as a function of the length, and compute the cost of building 13.2 miles.

Solution   The first sentence tells us that $C = kL$, for some fixed k. The constant k can be computed from the information given in the second sentence, which can be rephrased symbolically as $1{,}000{,}000 = (k)(2.5)$.

$$\therefore k = 400{,}000, \text{ and } C = 400{,}000L.$$

Hence, the cost C of building 13.2 miles is

$$C = (400{,}000)(13.2) = 5{,}280{,}000, \text{ and so the cost is } \$5{,}280{,}000. \quad \blacksquare$$

Remarks

a)   The value of k in the example above depends on the units chosen for cost and length. If you choose dollars and feet, then k would be $\dfrac{C}{L} = \dfrac{1{,}000{,}000}{(2.5)(5280)} \cong 75.8$. It is important to be consistent in using units. The final cost, however, will be the same, no matter which units you choose.

b)   Once you have chosen the units you wish to use and have made sure that all quantities are consistently expressed in those units **only**, you can solve the problem without explicitly writing down the units at each step of the calculation. All calculated quantities will automatically come out in the chosen units.

EXAMPLE 6   The gravitational force between two point-masses satisfies the inverse square law with respect to the distance between them. Suppose the gravitational force acting on you at sea level is 150 lbs, and that you and the earth can be considered point-masses with the mass concentrated at the centers. If the radius of the earth is approximately 4000 mi, and Mount Everest is 29,028 ft high, estimate the gravitational force acting on you on top of Mount Everest.

Solution   First of all, 29,028 ft $\cong$ 5.5 mi. We know the force F satisfies the inverse square law with respect to distance, so

$$F = \frac{k}{r^2},$$

where k is some constant, depending on you and the earth, and $r =$ the distance between you and the center of the earth.

At sea level, you have

$$50 = \frac{k}{4000^2},$$

and so $k = 2{,}400{,}000{,}000 = 2.4 \times 10^9$, and hence

$$F = \frac{2.4 \times 10^9}{r^2}.$$

On top of Mount Everest, $r = 4000 + 5.5 = 4005.5$, and so

$$F = \frac{2.4 \times 10^9}{(4005.5)^2} = \frac{2.4 \times 10^9}{(4.0055)^2 \times 10^6} \cong 0.1496 \times 10^3 = 149.6 \text{ lb.}$$

So the force of gravity pulls you almost half a pound less than it would at sea level, for example, in Boston. ■

EXAMPLE 7 Bob drives his pickup truck at 60 mph for a 100-mi trip. Brenda figures she can do the same trip faster in her little deuce coupe.

a) If they need to arrive at the same time, express the speed S at which she needs to drive as a function of the time W she waits to leave after Bob has already left.

b) If she waits half an hour, how fast will she have to drive?

Solution a) This is a distance-rate-time problem. The basic equation is:

$$\text{rate} = \frac{\text{distance}}{\text{time}}, \text{ or } R = \frac{D}{T},$$

and so $D = R \cdot T$ and $T = \frac{D}{R}$. It is useful to make a chart and to fill in what we know (using miles and hours):

	Bob's Trip	Brenda's Trip
D	100 (given)	100 (given)
R	60 (given)	S (given)
T	$\frac{100}{60}\left(T = \frac{D}{R}\right)$	$\frac{100}{S}\left(T = \frac{D}{R}\right)$

Notice that the total time is $\frac{100}{60}$ for Bob's trip and $W + \frac{100}{S}$ for Brenda's wait-plus-trip. So $\frac{100}{60} = W + \frac{100}{S}$. Solving for S will give the result:

$$\frac{100}{S} = \frac{100}{60} - W = \frac{5}{3} - W = \frac{5 - 3W}{3}$$

$$\frac{100}{S} = \frac{5 - 3W}{3}$$

$$\frac{S}{100} = \frac{3}{5 - 3W}$$

$$S = \frac{300}{5 - 3W} .$$

b) If $W = \frac{1}{2}$, then $S = \dfrac{300}{5 - \dfrac{3}{2}} = \dfrac{300}{3.5} \cong 86 \text{ mph.}$

(Maybe it's better to leave a little earlier!) ■

10.1 Exercises

1) A rectangular field of area 20,000 sq ft is to be fenced off next to a river, with fencing on three sides and the river making the fourth side. Express the length of fencing necessary as a function of the width of the field.

2) A rectangular field of area 10,000 sq ft next to a river is to be subdivided, into three congruent pens, as shown in the diagram. Assume the river makes the fourth side. Express the length of fencing necessary as a function of the width of the field, if no fencing is required at the river's edge.

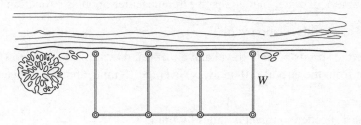

3) The shape of a window is given by two squares, one on top of the other, with a semicircle on top of that. Find the perimeter and area of the window as a function of the width of the window.

4) A cylindrical can is to be made up from sheet steel so that its surface area is 100 sq in. Express the height as a function of the radius. (*Hint*: Imagine removing the top and bottom with a can opener and splitting the rest down the side and unrolling it flat.)

5) An open cardboard box is to be constructed from a rectangular 8 in. by 10 in. sheet by cutting identical squares of side x out of each corner and folding up the resulting edges. Determine the volume of the box as a function of x.

6) An open cardboard box is to be constructed from a rectangular 1.5 m by 2 m sheet by cutting identical squares of side x out of each corner and folding up the resulting edges. Determine the volume of the box as a function of x. Also, what is the volume of the box as a function of the exterior surface area?

7) The volume of a sphere is $V = \frac{4}{3}\pi r^3$, and its surface area is $S = 4\pi r^2$. Express V as a function of S and vice versa.

8) A farm silo has the shape of a hemisphere on top of a cylindrical can of radius r and height h. If the total volume is 10,000 cubic meters, express the height as a function of the radius.

9) Suppose the gravitational force acting on you at sea level is 140 lb and that you and the earth can be considered point-masses with the mass concentrated at the centers. If the radius of the earth is approximately 4000 mi and the Matterhorn is 14,688 ft high, estimate the gravitational force acting on you on top of the mountain.

10) The distance between the earth and the moon varies from a maximum of approximately 253,000 mi to a minimum of around 221,000 mi. Find the percentage increase in the gravitational force acting between the two bodies when going from farthest separation to closest separation.

11) A 2-ft wire is cut into two pieces. One of the pieces, of length x, is bent into a circle. The other piece is bent into a rectangle whose length is twice the size of its width. What is the total area of the two shapes as a function of x?

12) Suppose you invest $2000 every year in a money market account that earns 5% a year simple interest. Write down an expression for how much money you have t years after you started investing. How much money will you have after 30 years?

13) When a tortoise crosses a highway, its speed is 2 ft/hr faster than normal. If it can cross a 24-ft lane in 24 min less time than it can travel that distance off the highway, then what is its normal speed?

14) A water tank contains 200 gal. of water. At 11 AM a pump starts to pump in more water at a rate of 5 gpm (gal. per min). At noon the tank springs a bad leak, losing water at 10 gpm.

   a) Express the volume V of water in the tank at t minutes after noon as a function of t.

   b) For what values of t is this relation valid?

15) We wish to compare the speeds of train travel and air travel, downtown to downtown. Assume that trains average 60 mph, taxis to and from the airport (10 mi away) average 15 mph, and planes average 300 mph with a 2-hr wait beforehand.

   a) Express the time T_t needed to travel m miles by train.

   b) Express the time T_a needed to travel m miles by air.

   c) For what value of m are they equal?

10.2 Solving Right Triangles

If you know certain sides and angles of a right triangle, the trigonometric functions can be used to find the rest, which is called **solving the triangle**. Since $\sin \theta = \frac{\text{OPP}}{\text{HYP}}$ if θ is in $\left(0, \frac{\pi}{2}\right)$ and if you know the size of θ, you can calculate (e.g., with a calculator or a table) the value of $\sin \theta$ and hence the ratio $\frac{\text{OPP}}{\text{HYP}}$. You can also get the value of the other five trigonometric functions. So, if you know θ and one of the sides, you can get the other sides also. Of course, you will make full use of the *Pythagorean theorem*: in a right triangle, the square of the hypotenuse is equal to the sum of the squares of the other two sides. This can be expressed as $(\text{HYP})^2 = (\text{OPP})^2 + (\text{ADJ})^2$.

EXAMPLE 1 Solve

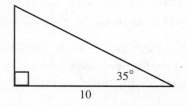

Solution We know that ADJ = side adjacent to the 35° angle = 10. To get OPP, the side opposite the angle, consider that in this case $\frac{\text{OPP}}{\text{ADJ}} = tan\ 35°$, and so OPP = ADJ · tan 35° ≅ (10) · (.7002) = 7.002. (Don't forget, when using your calculator, to make sure it is set on degrees when using degrees, not radians.)

Given the opposite side, OPP, you can get the hypotenuse, HYP, from the Pythagorean theorem, or by using $\frac{\text{HYP}}{\text{ADJ}} = sec\ 35°$, to obtain HYP = ADJ · sec 35° ≅ (10) · (1.2208) = 12.208. Of course, the last angle is 55°. (Right???) ■

EXAMPLE 2 A rocket blasts off vertically, its path being followed by a camera on the ground 2 miles away. Find a relation between the height of the rocket and the angle of elevation of the camera.

Solution First draw a picture and label the physical quantities. Let the variable h denote the height, or altitude, of the rocket in miles, and let the angle of elevation of the camera be θ.

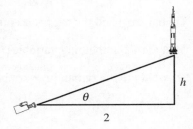

Notice that OPP = h and ADJ = 2, so that $\tan \theta = \frac{h}{2}$ and $h = 2 \tan \theta$. ■

EXAMPLE 3 Given the triangle below, find α, β, and x.

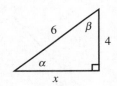

Solution Finding x is easy: $x = \sqrt{36 - 16} = \sqrt{20} = 2\sqrt{5}$.

To find α, notice that $\sin \alpha = \frac{4}{6} = \frac{2}{3} \cong .667$.

This problem is different from what you're used to. Typically, you are given an angle and then asked for the trigonometric value. Here you are given the trigonometric value and asked for the angle. You will learn about this further when you study inverse trigonometric functions. For now, however, you can solve the problem with your calculator. By entering the number .667 and then pressing the $\sin^{-1}$ key, you get $\alpha \cong 42°$, which means that $\beta \cong 48°$. ■

Remark Your calculator works in both degrees and radians (usually indicated as "Deg" and "Rad"). The default position depends on the particular model of calculator. Make sure you know which you are using!

10.2 Exercises

Consider the right triangle below, and use a calculator to find the angles (to the nearest degree) and the sides (to one decimal place) for Exercises 1–7.

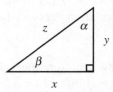

1) If $x = 3$ and $y = 5$, what are z and the angles α and β?

2) If $x = 3.01$ and $y = 2.75$, what are z and the angles α and β?

3) If $\alpha = 40°$ and $x = 10$, what are the remaining variables?

4) If $\alpha = 22.3°$ and $x = 12.1$, what are the remaining variables?

5) If $\alpha = 32°$ and $y = 8$, solve the triangle.

6) If $\alpha = 41°$ and $y = 10$, solve the triangle.

7) If $\beta = 54°$ and $x = 5$, find the rest.

8) Solve in terms of h:

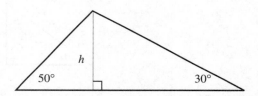

9) A fire ladder is mounted on its truck at a height of 4 ft. The ladder is 48-ft long, and it can rise at most 70° above the horizontal. If the first floor is street level and each floor is 9 ft high, assess your chances if you're standing on the sixth floor, looking out the window, in a hurry.

10) You are paddling in your kayak, exploring the bays of Alaska. A snowy mountaintop glistens in the morning sun. Your rangefinder shows that the top of the mountain is 10,000 ft diagonally away from you, and your sextant shows that the top is 25° above the horizon. How high is the mountain?

11) You can walk at 4 mph and row a boat at 2 mph. You are at point *A*, and need to cross a 2-mile wide river and get to point *B*, by rowing from *A* to the shore as shown on the following page, and then walking along the shore to *B*. Point *B* is 3 miles away from the point directly across the river from *A*. (Assume that the speed of the current is negligible.)

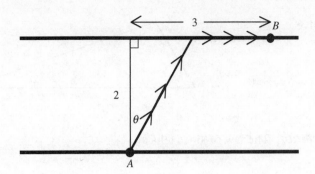

Express the time it will take you as a function of the angle θ. (Recall that $D = R \cdot T$, so $T = \frac{D}{R}$ and $R = \frac{D}{T}$.)

12)   For reasons of safety, the angle that a ladder makes with the ground should be no higher than 65°. You need the ladder to reach at least to the top of your 35-ft high house. How long is the shortest ladder you can use?

10.3 The Law of Sines and the Law of Cosines

Consider **any triangle** whose angles are A, B, and C, and whose sides opposite those angles are a, b, and c, respectively.

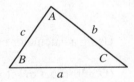

Law of Sines: $\dfrac{\sin A}{a} = \dfrac{\sin B}{b} = \dfrac{\sin C}{c}$.

Law of Cosines: $c^2 = a^2 + b^2 - 2ab \cos C$,

which also means that

$$a^2 = b^2 + c^2 - 2bc \cos A$$

and
$$b^2 = a^2 + c^2 - 2ac \cos B.$$

In the law of cosines, what happens if the angle is 90°? Does the equation look familiar? These laws are useful for solving acute or scalene triangles.

EXAMPLE 1 Solve

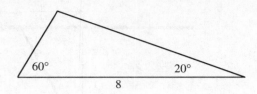

Solution The law of sines tells all.

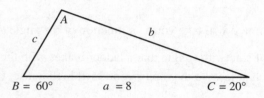

Notice that $A = 100°$, so

$$\frac{\sin 100°}{8} = \frac{\sin 60°}{b} = \frac{\sin 20°}{c}.$$

You can solve for b and c as

$$b = \frac{8 \sin 60°}{\sin 100°} \quad \text{and} \quad c = \frac{8 \sin 20°}{\sin 100°}.$$

Using a calculator or table, you can get decimal equivalents.

(Don't forget to set your calculator to degrees rather than radians.) ■

EXAMPLE 2 Solve

Solution The law of cosines looks more promising here. First, label the triangle:

$$c = 3 \quad \overset{A}{\underset{B = 80°\quad a = 4}{\triangle}} \quad b \quad C$$

So, $b^2 = a^2 + c^2 - 2ac \cos B$

$$= 16 + 9 - 24 \cos 80°$$

$$\cong 20.8$$

$$\therefore b \cong 4.56.$$

Now you can find the angle C from the law of sines:

$$\frac{\sin B}{b} = \frac{\sin C}{c}, \text{ so } \frac{\sin 80°}{4.56} = \frac{\sin C}{3}$$

and

$$\sin C = 3 \cdot \frac{\sin 80°}{4.56} \cong .648.$$

$\therefore C =$ angle, between $0°$ and $90°$, whose sine is .648. Using a calculator we find out that $C \cong 40°$.

$$\therefore A \cong 60° \quad \blacksquare$$

10.3 Exercises

Consider the following triangle:

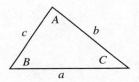

1) If $A = 100°$, $B = 50°$, and $b = 12$, find the rest.

2) If $a = 6$, $b = 5$, and $C = 60°$, solve the triangle.

3) If $a = 4$, $b = 5$, and $c = 6$, solve the triangle.

4) Let $C = 20°$, $c = 2$, and $b = 5$. Find two triangles with these measures. Draw the triangles.

5) What results when the angle mentioned in the law of cosines is $90°$?

CHAPTER

Trigonometric Identities

11

The basic idea is that antidifferentiating $\tan^2 x$ is really tough until you realize the fact that $\tan^2 x = \sec^2 x - 1$, and then there's nothing to it, because you know how to integrate $\sec^2 x$ and 1. That's the point: change what you have into something that you can handle. Here are the basic tools from Chapter 7.

Basic Identities:

$$\tan x = \frac{\sin x}{\cos x} \qquad \cot x = \frac{\cos x}{\sin x}$$

$$\sec x = \frac{1}{\cos x} \qquad \csc x = \frac{1}{\sin x}$$

Circular Identities: $\boxed{\sin^2 x + \cos^2 x = 1}$, which implies

$$\sin^2 x = 1 - \cos^2 x, \text{ so } \sin x = \pm\sqrt{1 - \cos^2 x}$$

and

$$\cos^2 x = 1 - \sin^2 x, \text{ so } \cos x = \pm\sqrt{1 - \sin^2 x}.$$

Notice that for $\sin x = \pm\sqrt{1 - \cos^2 x}$, you can't say whether it is "+"or "−", unless you have some additional information, for example, the quadrant that x is in. If you divide both sides of the equation $\sin^2 x + \cos^2 x = 1$ by $\cos^2 x$, you get

$$\boxed{\tan^2 x + 1 = \sec^2 x},$$

which gives

$$\tan^2 x = \sec^2 x - 1.$$

Taking square roots gives the following equations:

$$\tan x = \pm\sqrt{\sec^2 x - 1}$$

and

$$\sec x = \pm\sqrt{\tan^2 x + 1}.$$

Again the sign may be determined if you have extra information. Lastly, dividing both sides of the equation $\sin^2 x + \cos^2 x = 1$ by $\sin^2 x$ gives

$$\boxed{1 + \cot^2 x = \csc^2 x}\,,$$

which gives $\qquad\qquad \cot^2 x = \csc^2 x - 1.$

Taking square roots now gives

$$\cot x = \pm\sqrt{\csc^2 x - 1},$$

and $\qquad\qquad\qquad \csc x = \pm\sqrt{1 + \cot^2 x}.$

We also have the **Addition Formulas** (MUST BE MEMORIZED):

$$\boxed{\sin(A + B) = \sin A \cos B + \cos A \sin B}$$

$$\boxed{\cos(A + B) = \cos A \cos B - \sin A \sin B}$$

These formulas have several often-used consequences:

1) $\quad \sin 2x = \sin(x + x)$
$$= \sin x \cos x + \cos x \sin x$$
$$\boxed{\sin 2x = 2 \sin x \cos x}$$

2) $\quad \cos 2x = \cos(x + x)$
$$= \cos x \cos x - \sin x \sin x$$
$$\boxed{\cos 2x = \cos^2 x - \sin^2 x}$$

3) $\quad$ Since $\cos 2x = \cos^2 x - \sin^2 x$, we have
$\cos 2x = (1 - \sin^2 x) - \sin^2 x = 1 - 2\sin^2 x$, which gives

$$\boxed{\sin^2 x = \frac{1 - \cos 2x}{2}} \quad \text{and so} \quad \boxed{\sin x = \pm\sqrt{\frac{1 - \cos 2x}{2}}}$$

Also, since $\cos 2x = \cos^2 x - \sin^2 x$, we have
$\cos 2x = \cos^2 x - (1 - \cos^2 x) = 2\cos^2 x - 1$, which gives:

$$\boxed{\cos^2 x = \frac{1 + \cos 2x}{2}} \quad \text{and so} \quad \boxed{\cos x = \pm\sqrt{\frac{1 + \cos 2x}{2}}}\,.$$

The equations involving $\sin^2 x$ and $\cos^2 x$ will come in handy when doing antidifferentiation. The other two equations with the square roots provide a nice way to determine sine and cosine of an angle, knowing the cosine of twice the angle. Finally, by substituting $-B$ for B into the addition formulas, you get the difference formulas.

4) Since $\sin(A - B) = \sin(A + (-B))$,

$$\sin(A - B) = (\sin A)(\cos(-B)) + (\cos A)(\sin(-B))$$
$$= (\sin A)(\cos B) + (\cos A)(-\sin B)$$

or $\quad \boxed{\sin(A - B) = \sin A \cos B - \cos A \sin B}$

(See remark below.)

5) Since $\cos(A - B) = \cos(A + (-B))$,

$$\cos(A - B) = (\cos A)(\cos(-B)) - (\sin A)(\sin(-B))$$
$$= (\cos A)(\cos B) - (\sin A)(-\sin B)$$

or $\quad \boxed{\cos(A - B) = \cos A \cos B + \sin A \sin B}$

(See remark below.)

Remark You can either memorize the results of parts (1) through (5) or recall that it all comes from the addition formulas, while remembering that for parts (4) and (5)

$$\boxed{\cos(-\theta) = \cos \theta}\,,$$

and $\quad \boxed{\sin(-\theta) = -\sin \theta}\,.$

How do we know that? Look at this diagram:

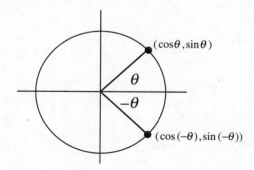

Notice that the first coordinates are the same and so $\cos(-\theta) = \cos \theta$. The second coordinates are negatives of each other, and so $\sin(-\theta) = -\sin \theta$.

EXAMPLE 1   Without using a calculator, determine $\sin 75°$.

Solution Since you can write $75° = 30° + 45°$,

$$\sin 75° = \sin(30° + 45°) = \sin 30° \cos 45° + \cos 30° \sin 45°$$

using the **Addition Formula**. Since $\sin 30° = \frac{1}{2}$, $\cos 30° = \frac{\sqrt{3}}{2}$, and $\sin 45° = \cos 45° = \frac{\sqrt{2}}{2}$ we have

$$\sin 75° = \frac{1}{2} \cdot \frac{\sqrt{2}}{2} + \frac{\sqrt{3}}{2} \cdot \frac{\sqrt{2}}{2} = \frac{\sqrt{2}}{4}(1 + \sqrt{3}). \quad \blacksquare$$

EXAMPLE 2   Without using a calculator, determine $\cos 15°$.

Solution Since $15° = 45° - 30°$, we have

$$\cos 15° = \cos(45° - 30°) = \cos 45° \cdot \cos 30° + \sin 45° \cdot \sin 30°$$

or $\qquad\qquad \cos 15° = \dfrac{\sqrt{2}}{2} \cdot \dfrac{\sqrt{3}}{2} + \dfrac{\sqrt{2}}{2} \cdot \dfrac{1}{2} = \dfrac{\sqrt{2}}{4}(\sqrt{3} + 1).$

Notice that this is the same answer as in the last example. $\quad \blacksquare$

If you use the subtraction formulas with $A = \frac{\pi}{2}$ and $B = \theta$, you have the following two identities:

$$\boxed{\sin\left(\frac{\pi}{2} - \theta\right) = \cos\theta} \quad \text{and} \quad \boxed{\cos\left(\frac{\pi}{2} - \theta\right) = \sin\theta}.$$

Note that given an angle θ, the **complementary angle** is $\frac{\pi}{2} - \theta$. Cosine is actually the complementary-sine function, since the sine of the angle equals the cosine of the complementary angle, and vice versa. Since $15 = 90 - 75$, $\sin 75°$ therefore equals $\cos 15°$.

Note You can write the addition and subtraction formulas together as

$$\boxed{\sin(A \pm B) = \sin A \cos B \pm \cos A \sin B} \qquad A$$

and $\qquad\qquad \boxed{\cos(A \pm B) = \cos A \cos B \mp \sin A \sin B}.$

11.1 Exercises

1) Write $\cos^7 x$ as $\cos x \cdot$ (some function of $\sin x$).

2) Write $\cos^5 x$ as $\cos x \cdot$ (some function of $\sin x$).

3) Write $\sec^5 x$ as $\sec^2 x \cdot$ (some function of $\tan x$).

4) Write $\sec^7 x$ as $\sec^2 x \cdot$ (some function of $\tan x$).

5) Write $\dfrac{\tan^5 x}{\cos x}$ as $(\sec x \tan x) \cdot$ (some function of $\sec x$).

6) Calculate $\cos 120°$ and $\sin 15°$ using the sum and/or difference formulas.

7) Use identities to show that $\tan x + \cot x = (\sec x)(\csc x)$.

8) Using the sum and difference formulas for sine and cosine, derive the following results:
$$\tan(A + B) = \frac{\tan A + \tan B}{1 - \tan A \tan B} \text{ and } \tan(A - B) = \frac{\tan A - \tan B}{1 + \tan A \tan B}.$$

9) Using the results in Exercise 8, write $\tan 2A$ in terms of $\tan A$.

10) Using the sum and difference formulas for sine and cosine, derive the following results:
$$\cot(A + B) = \frac{\cot A \cot B - 1}{\cot B + \cot A} \text{ and } \cot(A - B) = \frac{\cot A \cot B + 1}{\cot B - \cot A}.$$

11) Using the results in exercise 10, write $\cot 2A$ in terms of $\cot A$.

12) Show that $\cot\left(\frac{\pi}{2} - \theta\right) = \tan \theta$.

13) Show that $\csc\left(\frac{\pi}{2} - \theta\right) = \sec \theta$.

14) a) Using the equation $\cos^2 x = \dfrac{1 + \cos 2x}{2}$, write $\cos^4 x$ in terms of $\cos 2x$ and $\cos^2 2x$.

    b) Now use the equation $\cos^2 2x = \dfrac{1 + \cos 4x}{2}$ to write $\cos^4 x$ in terms of $\cos 2x$ and $\cos 4x$ to the **first** power.

15) a) Using the equation $\sin^2 x = \dfrac{1 - \cos 2x}{2}$, write $\sin^4 x$ in terms of $\cos 2x$ and $\cos^2 2x$.

    b) Now use the equation $\cos^2 x = \dfrac{1 + \cos 2x}{2}$ $\left(\text{or rather use the form } \cos^2 2x = \dfrac{1 + \cos 4x}{2}\right)$ to write $\sin^4 x$ in terms of $\cos 2x$ and $\cos 4x$ to the **first** power.

Exponential Functions

A.1 Introduction

An exponential function has the form $f(x) = a^x$, where $a > 0$. The number a is called the **base**. Consider $a = 2$. It is clear what $f(x) = 2^x$ means for some values of x. For example,

$$f(0) = 2^0 = 1, \qquad\qquad f(1) = 2^1 = 2,$$
$$f(2) = 2^2 = 4 \qquad\qquad f(3) = 2^3 = 8,$$
$$f(-1) = 2^{-1} = \frac{1}{2}, \qquad\qquad f(-2) = 2^{-2} = \frac{1}{4},$$
$$f\left(\frac{1}{2}\right) = 2^{1/2} = \sqrt{2} \cong 1.414,$$

and
$$f(3.2) = 2^{3.2} = 2^3 \cdot 2^{1/5} = 8\sqrt[5]{2}.$$

This last one could be tough to calculate, but at least you know **what it means.** Use the above values to plot these points for the graph of $y(x) = 2^x$, shown in Figure A.1.

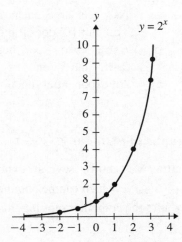

Figure A.1

We know what all rational exponents of 2 mean: $2^{m/n} = \left(\sqrt[n]{2}\right)^m$, if $\frac{m}{n}$ is in lowest terms. What happens at all the remaining (irrational) points? It can be shown (but not here) that there is exactly **one** smooth curve, always increasing, that can be drawn through all these points. That is the graph of 2^x, for x real. Notice that its domain is $(-\infty, \infty)$ and the range is $(0, \infty)$.

Graphs of a^x, for $a > 1$

Using these methods, plot the family of graphs a^x. Figure A.2 shows several exponential functions for the case when $a > 1$.

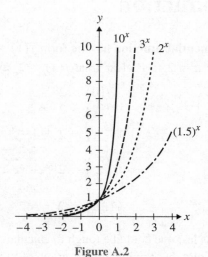

Figure A.2

Notes

a) As x gets large, each of these functions increases without bound (goes to ∞), but 10^x does it much faster than 3^x, etc.

b) As x goes to $-\infty$, these functions go to 0. But again, 10^x does it much faster than 3^x, etc.

c) All exponential functions pass through the point (0, 1).

Graphs of a^x, for $0 < a < 1$

Figure A.3 shows several exponential functions for the second case, where $0 < a < 1$, which can be obtained by plotting a few points. (A few thousand that is, if you're the computer plotting this graph.)

By comparing Figures A.2 and A.3, it certainly **looks** as if $\left(\frac{1}{10}\right)^x$ is the mirror image of 10^x, and $\left(\frac{1}{2}\right)^x$ is the mirror image of 2^x, etc. That this is true can be seen by a little computation. Let $f(x) = 2^x$, and let $g(x) = \left(\dfrac{1}{2}\right)^x$. (Remember: $g(x)$ is the mirror image about the y-axis of $f(x)$ if and only if $g(x) = f(-x)$.) Well, $g(x) = \left(\dfrac{1}{2}\right)^x = \dfrac{1^x}{2^x} = \dfrac{1}{2^x} = 2^{-x} = f(-x)$. This point is made graphically in Figure A.4.

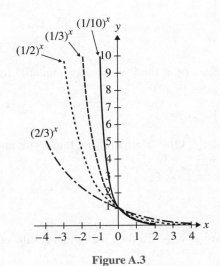

Figure A.3

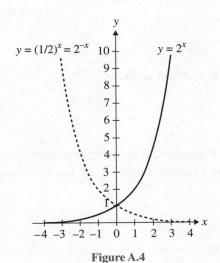

Figure A.4

By the way, notice that a^x is defined **for all x** only if $a > 0$. If $a < 0$, you can no longer have a^x for all x. For example if $a = -1$, and $x = \frac{1}{2}$, we have $(-1)^{1/2} = \sqrt{-1}$, which is not a real number. So the family of functions a^x is defined for $a > 0$ and any real number x.

A.1 Exercises

1) Determine the behavior of the following exponential functions as $x \to \pm\infty$, then sketch the graph of the function, labeling at least three points:

   a) $f(x) = \left(\dfrac{2}{3}\right)^x$

   b) $f(x) = \left(\dfrac{3}{2}\right)^x$

   c) $f(x) = 1.1^x$

   d) $f(x) = 0.32^x$

2) Determine the behavior of the following exponential functions as $x \to \pm\infty$, then sketch the graph of the function, labeling at least three points:

   a) $f(x) = \left(\dfrac{1}{3}\right)^x$

   b) $f(x) = \left(\dfrac{5}{2}\right)^x$

3) Graph the function $f(x) = 2^x$, then using that result and the methods learned in Chapters 4 and 7, graph the following functions:

   a) $f(x) = 2^{x-1}$

   b) $f(x) = 2^{x+3}$

   c) $g(x) = -2^{x+2}$

   d) $f(x) = -2^{x-1}$

4) Graph the function $f(x) = 3^x$, then using that result, graph the following functions:

   a) $f(x) = 3^{x+1}$

   b) $f(x) = -\frac{1}{2} \cdot 3^{x+1}$

5) If $f(x) = \cos x$ and $g(x) = 3^x$, find:

   a) $f(g(x))$

   b) $g(f(x))$

   c) $f(g(f(x)))$

6) If $f(x) = 2^x$ and $g(x) = \sin x$, find:

   a) $f(g(x))$

   b) $g(f(x))$

   c) $f(g(f(x)))$

7) Use the methods of Section 8.2 to decompose the following functions (i.e., find $f(x)$ and $g(x)$ so that the given function is equal to $f(g(x))$, where $f(x)$ is the outermost function):

   a) 2^{x-1}

   b) $3^{\sin x}$

   c) $5^{\cos(x^2)}$

8) Decompose the functions (see above):

   a) $\sin(3^x)$

   b) $\sqrt{3^{x+2}}$

   c) $2^{\sqrt{\sin x}}$

A.2 The Function e^x *(Also Called "The" Exponential)*

Of all the exponential functions a^x, the one that has a 45° tangent at $x = 0$ is especially important. This exponential is depicted in Figure A.5.

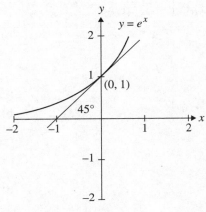

Figure A.5

You can tell from the graphs of 2^x and 3^x that this special function lies between them. The particular value of a that gives this exponential is called e. It can be calculated that $e \cong 2.718$. (In fact, $e \cong 2.718281828459045 \ldots$.) The exact reason why e^x is so important becomes clear in calculus, when you see that of all a^x, e^x has the simplest derivative.

Since any function a^x can be written as e^{kx}, for some k, that is the form ordinarily used in mathematics, science, and engineering. (See Appendix C for details.)

A.2 Exercises

1) Sketch:

   a) $y = e^{x-1}$ b) $y = e^{-x}$

   c) $y = -\frac{1}{2}e^x$ d) $f(x) = e^x - 1$

   e) $y = -\frac{1}{2}e^{-x}$

2) Sketch:

   a) $y = -e^{-x}$ b) $y = -2e^{-x}$

   c) $y = e^{-x} + 1$ d) $y = 3 - e^x$

   e) $y = 2 - 3e^x$

3) Sketch e^{-x^2}. (*Hint*: Plot and think!)

4) Sketch $y = e^x + e^{-x}$.

5) Let $f(x) = e^x$ and $g(x) = \tan x$. Find:

    a) $f(g(x))$ b) $g(f(x))$

    c) $\cos(f(f(x)))$

6) Let $f(x) = \sin x$ and $g(x) = e^{-x}$. Find:

    a) $f(g(x))$ b) $g(f(x))$

    c) $g(g(x))$

7) Simplify:

    a) $(e^x)^3$ b) $\dfrac{e^{2x}}{e^x}$

    c) $\dfrac{e^{2x} - 1}{e^x - 1}$

8) Simplify:

    a) $(e^{-x})^2$ b) $\sqrt{e^{2x}}$

    c) $\dfrac{e^x + 1}{e^{2x} - 1}$

9) Use the methods of Section 8.2 to decompose the following functions (i.e., find $f(x)$ and $g(x)$ so that the given function is equal to $f(g(x))$, where $f(x)$ is the outermost function):

    a) e^{3x+1} b) $\cos(e^x)$

    c) $\sin(e^{x^2})$ d) $\dfrac{1}{\cos(e^x)}$

    e) $\cos^2(e^x)$

10) Decompose the following:

    a) e^{-x} b) $\tan(e^{-x})$

    c) $e^{-\tan x}$ d) $\sqrt{e^{\cos x}}$

    e) e^{e^x}

11) Sketch $e^{-x} \cos x$.

12) Suppose a bacterial culture is growing such that it doubles its mass every 2 hr. If the culture begins as a mass of 1 g, write down a function that gives the mass at any time t. Graph this function. Under these conditions, what happens to the mass of the culture as $t \to \infty$?

13) Use your calculator to find:

    a) $e^{2.15}$ b) e^{-10}

    c) the value of x for which $e^x = 5$

14) Use your calculator to find:

    a) e^3 b) the value of x for which $e^x = 0.1$

    c) the value of x for which $2e^x + \sin x = 1$

Inverse Functions

B.1 ## The Idea of Inverses

Before discussing inverses, we remark that there are many different ways of talking about functions and their operations. For example, consider the function $f(x) = x^3$ at the point $x = 2$. Then all of the following ways of speaking mean the same thing.

1) Taking f of 2 gives you 8.
2) Evaluating f at 2 yields 8.
3) f operating on 2 gives 8.
4) Applying f to 2 gives 8.
5) f takes 2 to 8.
6) f acting on 2 yields 8.
7) $f(2) = 8$.
8) $2 \overset{f}{\longmapsto} 8$
9) $f: 2 \longmapsto 8$
10) $(2, 8)$ is a point on the graph of f.

All of these ways of saying the same thing are actually used. They all have their advantages in various different contexts.

When one function undoes the action of another, it is said to be the **inverse of the other**. For example, look at $f(x) = x^3$ and $g(x) = \sqrt[3]{x}$. If you take any number, x, cube it, and then take the cube root of the result, you're back to x. (Try this for $x = 2$ and 3.) In symbols

$$\sqrt[3]{x^3} = x \ \text{ or } \ g(f(x)) = x.$$

Similarly, you can show $\quad \left(\sqrt[3]{x}\right)^3 = x \ \text{ or } \ f(g(x)) = x.$

A second example is the doubling function, $f(x) = 2x$, and the halving function, $g(x) = \frac{x}{2}$; they're inverse to each other. The function $f(x) = \dfrac{1}{x}$ is its own inverse!

DEFINITION

> We say $f(x)$ and $g(x)$ are **inverse to each other** if $f(g(x)) = x$ and $g(f(x)) = x$ and if domain of f = range of g and domain of g = range of f. The inverse of the function f is denoted f^{-1}.

Note that the domains and ranges are important to the discussion of inverse functions. Two function expressions can be inverses over one interval but not inverses over another interval. (See Exercise 8 in Section B.3.) The notation for the inverse, f^{-1}, is standard, but it is an unfortunate choice because it **looks** like f to the power -1. So, if you really want to say f to the power -1, you should write $\frac{1}{f}$.

Inverses, and finding them, are a big deal in mathematics. Here's just a little example. Suppose you wish to solve the equation $f(x) = 0$. If you could find f^{-1}, you could apply it to both sides to get $f^{-1}(f(x)) = f^{-1}(0)$, and so $x = f^{-1}(0)$. Notice that the left side reduces to x, so PRESTO! You've solved the equation.

The question is: Okay, suppose that you've got a function f; how do you find f^{-1}? Specifically: if you've got the graph of f, how do you find the graph of f^{-1}? Or, if you've got an expression for f, how do you get the expression for f^{-1}? Read on!

B.2 Finding the Inverse of f Given by a Graph

First of all, notice that the graph of f is the set of points of the form $(x, f(x))$. (Do you agree?) For every x in the domain of f, f takes x to $f(x)$. Use the following symbols:

$$x \xmapsto{\;f\;} f(x).$$

Notice that f^{-1} takes $f(x)$ to x. So:

$$f(x) \xmapsto{\;f^{-1}\;} x.$$

Hence the graph of f^{-1} is exactly the set of points $f(x), x)$. (Notice, for example, that the point $(2, 8)$ is on the graph of x^3, and $(8, 2)$ is on the graph of $\sqrt[3]{x}$.) The result is that the graph of f^{-1} is just the graph of f with the order of the coordinates reversed. How do you do that, you ask? The picture in Figure B.1 tells all!

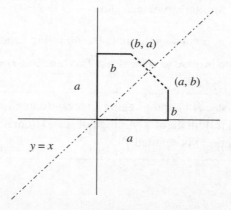

Figure B.1

Using geometry, you can prove that (*a*, *b*) and (*b*, *a*) are reflections of each other about the line $y = x$, that is, the line through the origin at an angle of 45°. **So to go from the graph of *f* to the graph of f^{-1}, simply reflect the entire graph of *f* about the line $y = x$.**

Figure B.2 shows the inverse functions $f(x) = 2x$ and $g(x) = \frac{x}{2}$.

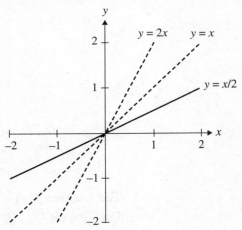

Figure B.2

Notice that each graph is obtained from the other by reflecting across the line $y = x$.

Consider the functions $f(x) = x^3$ and $g(x) = \sqrt[3]{x}$. They are also reflections of each other about the line $y = x$, as shown in Figure B.3.

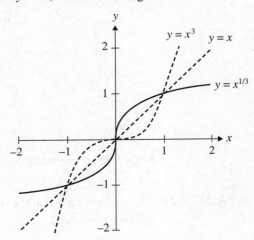

Figure B.3

EXAMPLE 1   Show that the functions $f(x) = x + 2$ and $g(x) = x - 2$ are inverses of one another on the interval $(-\infty, \infty)$, and then graph the functions.

Solution Since $f(g(x)) = f(x - 2) = (x - 2) + 2 = x$ and $g(f(x)) = g(x + 2) = (x + 2) - 2 = x$ for any value of x, these two functions are inverses on the entire interval $(-\infty, \infty)$. The graphs are shown below.

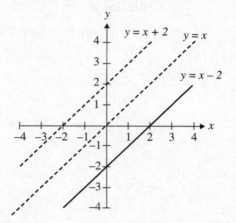

The function $f(x) = \frac{1}{x}$ is its own inverse since it reflects back onto itself, as shown in Figure B.4.

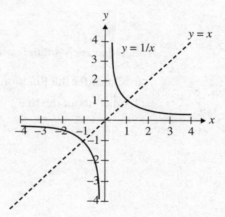

Figure B.4

Notice that some functions don't have inverse functions. Here's an example.

EXAMPLE 2 Show graphically that the function $f(x) = x^2$ does not have an inverse function on the interval $(-\infty, \infty)$.

Solution A picture's worth a thousand words.

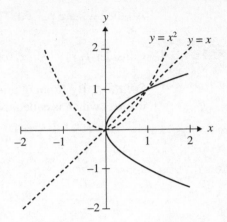

You can flip the graph of $f(x) = x^2$, but the resulting graph (given by the solid curve) is NOT a function. Think about it: how would you evaluate it at $x = 1$, for example? There are two candidates, 1 and -1. A function can have at most one value for each x. In other words, each vertical line can cross the graph of a function at most once. That's called the **vertical line test for being a function.** We conclude that $f(x) = x^2$ on $(-\infty, \infty)$ does not have an inverse function. ■

We got into this mess because our original function $f(x) = x^2$ had points where two different x values had the same $f(x)$ value. That is, it was crossed by horizontal lines more than once each. It flunked the **horizontal line test for a function to have an inverse.** Any function whose graph is crossed by a given horizontal line at most once is called **invertible**, that is, f^{-1} exists. Another term for invertible is **one-to-one,** written 1-1. (Why is this a good name?) So the problem was that $f(x) = x^2$ is not 1-1. But suppose you really wanted an inverse for $f(x) = x^2$. You could: a) get depressed or b) settle for a piece of the whole thing, as in the next example.

EXAMPLE 3   Consider the function $f(x) = x^2$, for $x \geq 0$. (Notice the restricted domain.) Graph its inverse.

Solution No problem!

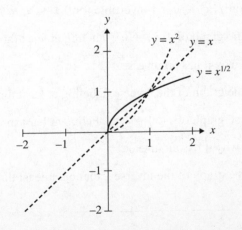

All the tests are passed. That inverse, by the way, is $g(x) = x^{1/2} = \sqrt{x}$. ■

EXAMPLE 4   Consider $f(x) = \sin x$, for $-\frac{\pi}{2} \le x \le \frac{\pi}{2}$. Graph its inverse.

Solution By flipping this restricted sine function about the line $y = x$, you'll get its inverse, which is called **inverse sine** and also **arcsine**. It is denoted $\arcsin x$ or $\sin^{-1} x$.

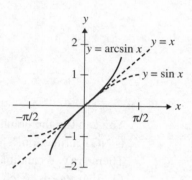

You'll see more of this in Appendix D. ■

B.2 Exercises

Graph Exercises 1–9, and their inverses, if they exist. If they do not, explain why.

1) $f(x) = 2x - 1$

2) $f(x) = x^5$

3) $g(x) = -x$

4) $k(y) = \cos y$

5) $h(x) = 2^x$

6) $L(t) = 1^t$

7) $f(x) = \left(\frac{1}{2}\right)^x$

8) $f(x) = (x - 3)^2$

9) $f(x) = x^3 - 1$

10) Is the function $f(x) = \cos x$ invertible for $0 \le x \le \pi$? How about for the interval $-\frac{\pi}{2} \le x \le \frac{\pi}{2}$? (Sketch!)

11) a) Sketch the graph of $f(x) = e^x$, labeling at least three points on the graph.

   b) Explain why it has an inverse.

   c) Sketch the graph of the inverse, labeling at least three points on the graph.

12) a) Sketch the graph of $f(x) = 2^x$, labeling at least three points on the graph.

   b) Explain why it has an inverse.

   c) Sketch the graph of the inverse, labeling at least three points on the graph.

B.3 Finding the Inverse of f Given by an Expression

Suppose we have $f(x)$ given by an expression, in other words we have the equation $y = f(x)$, and suppose we want an expression for $f^{-1}(x)$. If we know f is invertible then f^{-1} exists, even though we don't yet know what it looks like. In any event, if we apply f^{-1} to both sides, we get

$$f^{-1}(y) = f^{-1}(f(x)),$$

and so
$$f^{-1}(y) = x.$$

But if you solved the equation $y = f(x)$ for x, you would get $x =$ some function of y. Hence $f^{-1}(y)$ is that function, meaning that to find $f^{-1}(y)$ you just solve the equation $y = f(x)$ for x. So here's the complete method for finding $f^{-1}(x)$.

1) Write down the equation $y = f(x)$.
2) Solve for x. (This gives you $x = f^{-1}(y)$.)
3) Take the expression $f^{-1}(y)$. Pluck out the y, put in the x, and voilà! You've got $f^{-1}(x)$.

EXAMPLE 1 Find the inverse of the function $f(x) = \dfrac{2}{x + 3}$.

Solution Write $y = \dfrac{2}{x + 3}$ and solve this equation for x to obtain $f^{-1}(y)$. Multiplying by $x + 3$ gives:

$$xy + 3y = 2,$$

which in turn means
$$xy = 2 - 3y,$$

and hence
$$x = \frac{2 - 3y}{y} = f^{-1}(y).$$

Using x instead of y gives

$$f^{-1}(x) = \frac{2 - 3x}{x}. \qquad \blacksquare$$

You can check this result by determining whether $f(f^{-1}(x)) = x$ and $f^{-1}(f(x)) = x$. In this case:

$$f^{-1}(f(x)) = f^{-1}\left(\frac{2}{x+3}\right)$$

$$= \frac{2 - 3\left(\dfrac{2}{x+3}\right)}{\dfrac{2}{x+3}} = \frac{\dfrac{2x+6-6}{x+3}}{\dfrac{2}{x+3}} = \frac{\dfrac{2x}{x+3}}{\dfrac{2}{x+3}} = \frac{2x}{2} = x.$$

Also $$f(f^{-1}(x)) = f\left(\frac{2-3x}{x}\right)$$

$$= \frac{2}{\dfrac{2-3x}{x}+3} = \frac{2}{\dfrac{2-3x+3x}{x}} = \frac{2}{\dfrac{2}{x}} = x.$$

Check! We're done.

EXAMPLE 2   Find the inverse of $f(x) = \sqrt[3]{2x+1}$.

Solution Write $y = \sqrt[3]{2x+1}$, and solve for x.

$$y^3 = 2x + 1$$

$$2x = y^3 - 1$$

$$x = \frac{y^3 - 1}{2} = f^{-1}(y),$$

$$\therefore f^{-1}(x) = \frac{x^3 - 1}{2} \quad \blacksquare$$

Question: It all looks so easy. Can anything go wrong? And what if a given function doesn't **have** an inverse? How will that show up?

Answer: Yes, things can deteriorate. For example, what if you can't solve for x? If that happens, then you're stuck. Moreover, if f^{-1} doesn't exist, it will show up by the fact that the equation **cannot** be solved for x. Not even by Gauss, with help from Einstein. See the next example.

EXAMPLE 3   Find the inverse of $f(x) = x^4 - 3$.

Solution Write $y = x^4 - 3$, and solve for x.

$$x^4 = y + 3$$

$$x = \pm\sqrt[4]{y + 3}$$

Aha, you see, that's not a function. You have not solved for x as a function of y. Of course, you knew that $f(x)$ does not have an inverse, because it flunks the horizontal line test.   ■

B.3   Exercises

In Exercises 1–6, find inverses, if they exist, of the given functions. If they do not, explain why.

1) $f(x) = 2x - 3$

2) $k(x) = \dfrac{x}{x + 1}$

3) $g(x) = \sqrt[3]{5x + 1}$

4) $s(t) = \sqrt{t + 2}$

5) $f(x) = \dfrac{2}{x}$

6) $f(w) = \dfrac{w^2}{w^2 + 1}$

7) The function $f(x) = (x - 1)^4$ does not have an inverse on the interval $(-\infty, \infty)$. Show this. Then show that, if you restrict the domain to $[1, \infty)$, this restricted function has an inverse. Graph both functions.

8) a) Let $f(x) = x^2 + 3$ on $[0, 1]$. Find its domain and range and then sketch it and its inverse. Find an expression for the inverse $f^{-1}(x)$.

   b) Let $g(x) = x^2 + 3$ on $[-1, 0]$. Do the same as in Part a).

   c) The function expressions of f and g are the same. Are their inverses the same?

9) The accompanying tables represent a function f that converts yards to feet and a function g that converts miles to yards.

x (yd)	1760	3520	5280	7040	8800
$f(x)$ (ft)	5280	10,560	15,840	21,120	26,400

x (mi)	1	2	3	4	5
$g(x)$ (yd)	1760	3520	5280	7040	8800

Evaluate each expression and interpret the results:

   a) $(f \circ g)(3)$

   b) $f^{-1}(21{,}120)$ (Note: 21,120 is not a pair of coordinates; it is simply the number 21,120.)

   c) $g^{-1}(3520)$

   d) $(g^{-1} \circ f^{-1})(26{,}400)$

Logarithmic Functions

C.1 Definition of Logarithms

Here's one more function you'll meet in science, engineering, and economics, and sometimes in the newspaper. In symbols, it looks like this:

$$\log_a x$$

It is read as **log, to the base *a*, of *x*.** (Do NOT read this as "the log of *a*-to-the-*x*"!) There are several ways of defining it. Here's one.

DEFINITION

> Let $a > 0$, $a \neq 1$. Then $\log_a x$ is **the number to which you raise *a* to get *x*.**

EXAMPLE 1 Demonstrate that $\log_2 8 = 3$.

Solution Here the base is 2 and $x = 8$. To what number do you have to raise 2 in order to get 8? Answer: 3, so $\log_2 8 = 3$. ■

EXAMPLE 2 Show that $\log_{10} 1,000,000 = 6$.

Solution Here the base is 10, and $x = 1,000,000$. What number do you have to raise 10 to in order to get 1,000,000 (6 zeros)? Answer: 6, so $\log_{10} 1,000,000 = 6$. ■

EXAMPLE 3 $\log_{10} .01 = ?$

Solution Here the base is 10, and $x = .01$. Write .01 as a power of 10. Here you go:

$$.01 = \frac{1}{100} = \frac{1}{10^2} = 10^{-2}$$

What number do you have to raise 10 to in order to get 10^{-2}? Answer: -2, of course, so $\log_{10} .01 = -2$. ■

EXAMPLE 4 $\log_2 32 = ?$

Solution Write 32 as a power of 2:

$$32 = 2 \cdot 2 \cdot 2 \cdot 2 \cdot 2 = 2^5,$$

so, $\log_2 32 = 5$. ■

EXAMPLE 5 $\log_3 81 = ?$

Solution Write 81 as a power of 3:

$$81 = 3 \cdot 3 \cdot 3 \cdot 3 = 3^4,$$

so, $\log_3 81 = 4$. ■

EXAMPLE 6 $\log_7 7^{15} = ?$

Solution 7^{15} is already written as a power of 7. What a silly question!
$\log_7 7^{15} = 15$. ■

EXAMPLE 7 $\log_a a^3 = ?$

Solution Another silly question! If you write a^3 as a power of a, obviously the exponent must be 3, so $\log_a a^3 = 3$. ■

Remark   The function $\log_{10} x$ is known as the **common logarithm**. Sometimes you will see the expression $\log x$. Usually that means $\log_{10} x$.

C.1 Exercises

Evaluate the following:

1) $\log_9 81$

2) $\log_{11} 121$

3) $\log_2 \dfrac{1}{2}$

4) $\log_3 \dfrac{1}{81}$

5) $\log \dfrac{1}{1000}$

6) $\log \dfrac{1}{100}$

7) $\log_{1/2} 8$

8) $\log_{1/4} 64$

9) $\log_3 \sqrt{3}$

10) $\log_7 \sqrt{7}$

11) $\log_4 2\sqrt{2}$

12) $\log_b b^{13}$

13) $\log_a a^x$

14) $2^{\log_2 8}$

15) $10^{\log_{10}\left(\frac{1}{100}\right)}$

16) $3^{\log_3 81}$

17) $3^{\log_3 80}$

18) $10^{\log_{10}(1000)}$

19) $10^{\log_{10}(15)}$

20) $10^{\log_{10} x}$

21) $a^{\log_a x}$

C.2 Logs as Inverses of Exponential Functions

(Graphs and Equations)

You will see that there is a fundamental relationship between logs and exponentials: they are inverse to each other. Recall from Appendix B that f and g are called inverse to each other if all of the following are true:

1) $f(g(x)) = x$
2) $g(f(x)) = x$
3) domain of f = range of g
4) domain of g = range of f

Theorem

Let $a > 0$, $a \neq 1$. Then $\log_a x$ and a^x are inverse to each other.

Remark This theorem can be proved rigorously using the continuity properties from calculus, but that cannot be done here. However, if you let $f(x) = a^x$ and $g(x) = \log_a x$, you can examine the first two conditions.

a) $f(g(x)) = a^{g(x)} = a^{\log_a x}$. What does this mean? It is a, raised to the number to which you raise a to get x. So it equals x, that is

$$f(g(x)) = a^{g(x)} = a^{\log_a x} = x.$$

b) $g(f(x)) = \log_a f(x) = \log_a a^x = x$. (Just like Examples 6 and 7 in the previous section!)

So you haven't completely proved this theorem, but you can see that $\log_a x$ and a^x undo each other.

Knowing that $\log_a x$ and a^x are inverses allows you immediately to graph $\log_a x$. If you wish to graph the function $f(x) = \log_2 x$, you need only graph the function $g(x) = 2^x$, and flip it around the line $y = x$ (see Figure C.1).

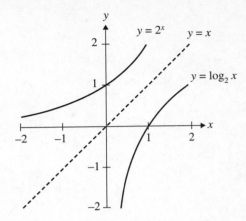

Figure C.1

Notice that the domain of $f(x) = \log_2 x$ is the set of all positive numbers, and the range is the set of all numbers. Notice also $\log_2 1 = 0$. Figure C.2 shows the common logarithm.

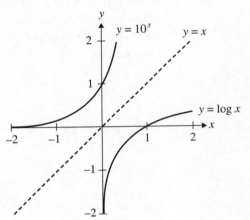

Figure C.2

The next example shows the logarithm for base $a = \frac{1}{2}$. First graph the function $\left(\frac{1}{2}\right)^x$, and then flip it about the line $y = x$ to obtain the graph of $f(x) = \log_{1/2} x$, as in Figure C.3.

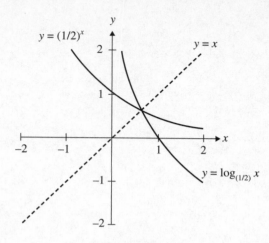

Figure C.3

Typically, you'll come across $\log_a x$ only for $a > 1$. Note that $\log_{1/a} x = -\log_a x$. You could get the previous graph by first graphing $\log_2 x$, and then rotating it about the x-axis! As with other functions, once you know what the log graph looks like, you can obtain other graphs by shifting and stretching. Figure C.4 shows how the graph of $\log_2(x + 3)$ is obtained from $\log_2 x$ by shifting it to the left three units.

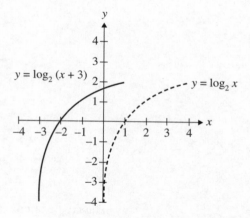

Figure C.4

By recalling the definition of logs and their relationship to exponentials, it is possible to solve some new equations containing such animals. The following examples show some of the problems you may encounter.

EXAMPLE 1   Solve $\log_2 x = 4$.

Solution We know that the function 2^x **undoes** the action of that log function. This means that if we apply the function 2^x to $\log_2 x$, we get x. [Recall that to apply the function $f(x)$ to the quantity a means to evaluate $f(x)$ at a, which gives us $f(a)$.]

Therefore, applying $f(x) = 2^x$ to $\log_2 x$ gives us $2^{\log_2 x}$, which we know is x, and applying $f(x) = 2^x$ to 4 gives us 2^4. So:

$$2^{\log_2 x} = 2^4$$

$$\therefore x = 16.$$

(The left side is x, because the log was "undone" by the action of 2^x; this is the true meaning of "inverseness," which is how $\log_2 x$ and 2^x are related.)   ■

EXAMPLE 2   Solve $\log_{10} x = 3$.

Solution As in the last example, to "undo" the action of the log (now with base 10), apply 10^x to both sides. (Recall that applying 10^x to 3 means evaluating 10^x at 3.)

So $x = 10^3 = 1000$.   ■

EXAMPLE 3   Solve $\log_{10}(x^2 - 4x + 14) = 1$.

Solution Again, apply 10^x to both sides. So,

$$(x^2 - 4x + 14) = 10^1 = 10,$$

or: $\qquad\qquad\qquad\qquad\qquad x^2 - 4x + 4 = 0,$

which factors as $\qquad\qquad (x - 2)^2 = 0.$ So $x = 2.$ ■

EXAMPLE 4   Solve $\log_3(x^2 - 3x - 7) = 1$.

Solution To "peel off" the log, apply 3^x, and in doing so you get:

$$(x^2 - 3x - 7) = 3^1 = 3,$$

or $\qquad\qquad\qquad\qquad\qquad x^2 - 3x - 10 = 0.$

Factor this last equation to give:

$$(x - 5)(x + 2) = 0,$$

which has solutions $x = 5$ and $x = -2$.   ■

EXAMPLE 5   Solve $2^{x^2+1} = 8$.

Solution Apply $\log_2 x$ to both sides (i.e., take $\log_2$ of both sides) and get

$$x^2 + 1 = \log_2 8,$$

or $$x^2 + 1 = 3.$$

This gives $$x^2 = 2,$$

or $$x = \pm\sqrt{2}. \quad \blacksquare$$

EXAMPLE 6 Solve $100^{\sin x} = 10$, for all $x \in \left[0, \dfrac{\pi}{2}\right]$.

Solution Apply $\log_{100} x$ to both sides.

Therefore $$\sin x = \log_{100} 10,$$

and since $$\log_{100} 10 = \frac{1}{2},$$

we have $$\sin x = \frac{1}{2},$$

or $$x = \frac{\pi}{6}. \quad \blacksquare$$

C.2 Exercises

1) a) Graph $f(x) = 2^{x+1}$.

   b) Show that $g(x) = \log_2 \dfrac{x}{2}$ is the inverse of $f(x)$ and graph it.

2) a) Graph $f(x) = 3^{x-1}$.

   b) Show that $g(x) = \log_3 3x$ is the inverse of $f(x)$ and graph it.

3) Solve the following:

   a) $2^{x-3} = 64$
   b) $3^{x+1} = 27$
   c) $4^{2x-3} = 16$
   d) $5^{x+5} = \dfrac{1}{125}$

4) Solve the following:

   a) $3^{x-2} = 27$
   b) $2^{x+1} = 64$
   c) $5^{2-x} = 125$

5) Solve $\log_3(x + 7) = -1$.

6) Solve $\log_3(x - 3) = 2$.

7)   Solve $\log_{64} x^2 = \dfrac{1}{3}$.

8)   Solve $\log_9 x^2 = \dfrac{1}{2}$.

9)   Solve $\log_3(x^2 - 5) = 2$.

10) Solve $\log(\cos x) = 0$ for $x \in [0, 2\pi]$.

11) Using a calculator, graph $y = \log x$ between 0.5 and 2 by using several x values. Using the same x values, graph $y = \log(10x)$. What do you notice?

C.3 Laws of Logarithms

Your life among the logs is made much simpler when you know certain log laws. They're great when you're solving equations or simplifying expressions.

Log Laws:

Let $a > 0$, $a \neq 1$, and let $x > 0$ and $y > 0$. Then:

1) $\log_a xy = \log_a x + \log_a y$

2) $\log_a \dfrac{x}{y} = \log_a x - \log_a y$

3) $\log_a x^r = r \log_a x$

4) $\log_a 1 = 0$ for all a (But we knew that already!)

5) $\log_a x = \dfrac{\log_b x}{\log_b a}$, for any convenient b, for example, 10 or e. This "change of base" law can be a lifesaver if you can't handle $\log_a x$.

EXAMPLE 1   Solve $\log_2 x^2 + \log_2 2x = 4$.

Solution You can combine the left side using rule 1.

So
$$\log_2 2x^3 = 4.$$

Now that the left side is a single log of an expression, you can apply 2^x to simplify it:
$$2x^3 = 2^4 = 16$$
$$x^3 = 8$$
$$x = 2.$$

You should check this solution:
$$\log_2(2^2) + \log_2(2 \cdot 2) = \log_2 4 + \log_2 4 = 2 + 2 = 4. \quad \blacksquare$$

EXAMPLE 2   Solve $\log_{10}(x^2 - 3x)^3 = 3$.

Solution Using the third law, you obtain

$$3 \log_{10}(x^2 - 3x) = 3,$$

or

$$\log_{10}(x^2 - 3x) = 1.$$

Now apply 10^x:

$$x^2 - 3x = 10^1 = 10.$$

Hence

$$x^2 - 3x - 10 = 0,$$

$$(x - 5)(x + 2) = 0,$$

and

$$x = 5 \text{ or } -2. \quad \blacksquare$$

In this example, you calculated two answers, both of which were valid because they satisfied the original equation. Sometimes, however, you may get some answers that are not valid. (They are called **extraneous solutions.** See Section 3.3) You must check your solutions to determine that they are valid. Consider the following.

EXAMPLE 3   Solve $\log_2 x + \log_2(x - 1) = 1$.

Solution Using the first law, you obtain

$\log_2 x + \log_2(x - 1) = \log_2(x(x - 1)) = \log_2(x^2 - x)$ which gives $\log_2(x^2 - x) = 1$. Now apply 2^x and get

$$x^2 - x = 2.$$

Hence

$$x^2 - x - 2 = 0$$

and

$$(x - 2)(x + 1) = 0,$$

and

$$x = 2 \text{ or } -1.$$

However, when checking these "solutions," notice that the initial equation is not satisfied for $x = -1$, so $x = 2$ is the only solution. $\blacksquare$

EXAMPLE 4   Evaluate $\log_7 5$.

Solution You can't evaluate this directly. You could use a calculator, but it doesn't have $\log_7 x$. But, using the change of base law,

$$\log_7 5 = \frac{\log_{10} 5}{\log_{10} 7}$$

$$\cong \frac{.699}{.845} \cong .827. \quad \blacksquare$$

C.3 Exercises

1) Solve $\log_3 x + \log_3(x - 6) = 3$.

2) Solve $\log_2 x + \log_2(x - 2) = 3$.

3) Solve $\log y + \log y^2 = -1$.

4) Solve $\log y^2 + \log y^3 = -3$.

5) Solve $\log_6(2x + 1) - \log_6(2x - 1) = 1$.

6) Solve $\log_2 x^2 - \log_2(3x - 8) = 2$.

7) Find numbers a, x, and y where the value of $\log_a(x + y)$ is not equal to the value of $\log_a x + \log_a y$. (*Hint:* Try some values of a, x, and y where these expressions are easy to calculate. Maybe you'll get lucky.)

8) Approximate $\log_3 4$ using the change of base formula and your calculator.

9) Solve $2 \log_2(\sin x) + 1 = 0$ for x in $[0, 2\pi]$.

10) Solve $\log x - \log(x - 1) - 1 = 0$.

11) Solve $10^{\sin x} = 1$ for x in $[0, 2\pi]$.

C.4 The Natural Logarithm

In Appendix A, we introduced the exponential function e^x. Its inverse, $\log_e x$, is called the **natural logarithm.** For simplicity, $\log_e x$ is denoted by the symbol $\ln x$. To get its graph, we flip e^x about the line $y = x$, as in Figure C.5.

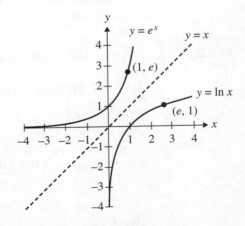

Figure C.5

Notice that $\ln 1 = 0$, and $\ln e = 1$. You will also encounter the natural logarithm when you study integration. It helps to know about $\ln x$, and how to graph related functions. Figure C.6 shows several horizontal shifts of the natural logarithm.

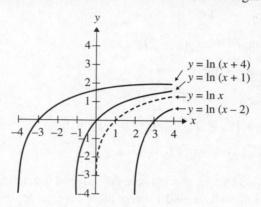

Figure C.6

In solving equations, you handle the natural log the same as any other log, remembering that the base is that special number e.

EXAMPLE 1   Solve $\ln(x^2 - 1)^3 = 1$.

Solution Apply e^x to get $(x^2 - 1)^3 = e$,

or
$$x^2 - 1 = e^{1/3},$$

which can be solved to give

$$x = \pm\sqrt{1 + e^{1/3}}.$$

For most purposes you can leave your answer in this form. If its decimal approximation is needed, you can use your calculator to determine one. ■

EXAMPLE 2   Solve $e^{\sin x} = 1, x \in [0, 2\pi]$.

Solution Apply $\ln x$ to get $\sin x = \ln 1 = 0$,

and so
$$x = 0, \pi, 2\pi. ■$$

Remark   As mentioned in Appendix A, any exponential a^x can be written in the form e^{kx} for some constant k. **THE TRICK:** write a as $e^{\ln a}$. So

$$a^x = \left(e^{\ln a}\right)^x = e^{(\ln a)x}$$

(Recall: $\left(a^m\right)^n = a^{mn}$.)

EXAMPLE 3   That's it! Now consider the following.

Write 2^x in the form e^{kx}.

Solution $2^x = \left(e^{\ln 2}\right)^x = e^{(\ln 2)x}$,

and since $\ln 2 \cong 0.693,$

$$2^x \cong e^{693x}. \quad \blacksquare$$

EXAMPLE 4   Write $(0.345)^x$ in the form e^{kx}.

Solution $(.345)^x = \left(e^{\ln .345}\right)^x = e^{(\ln .345)x}$,

and since $\ln 0.345 \cong -1.06,$

$$(0.345)^x \cong e^{-1.06x}. \quad \blacksquare$$

Remarks

a) If you convert the function a^x into the form e^{kx}, then if $a > 1$, as in
Example 3, the constant $k > 0$. All functions of the form e^{kx}, for $k > 0$,
look similar to the graph shown in Figure C.7.

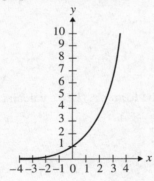

Figure C.7

In this case, the function is said to have **exponential growth.** An example
from biology would be the growth of a colony of bacteria under ideal conditions.

b) On the other hand, if $0 < a < 1$, as in Example 4, then the constant
$k < 0$. All functions of the form e^{kx}, for $k < 0$, look similar to the graph
shown in Figure C.8.

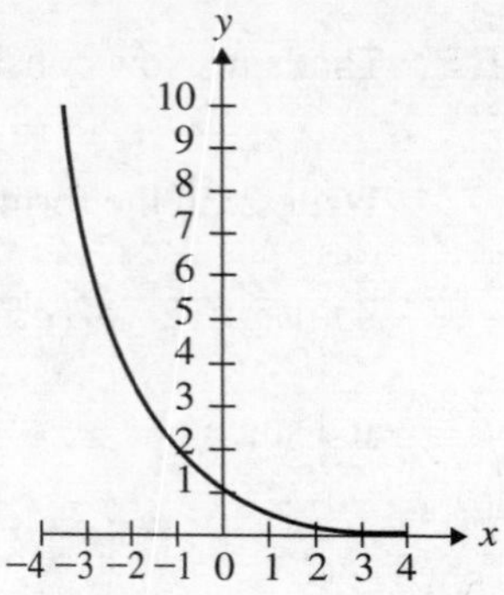

Figure C.8

In this case, the function is said to have **exponential decay.** In physics, the decay of radioactive substances is represented by such functions.

C.4 Exercises

1) Graph e^{x-1}.

2) Graph $e^{x+\pi}$.

3) Graph $y = \ln(x - 1)$.

4) Graph $y = 2 + \ln(x + 1)$.

5) Graph $y = 10 + \ln(x + 3)$.

6) Solve $\ln t - \ln t^2 = 5$.

7) Solve $\ln t + \ln t^2 = 6$.

8) Solve $e^{x^2+4x-5} = 1$.

9) Solve $e^{x^2+2x-3} = 1$.

10) Solve $e^{\ln(w^2+1)} = 5$. (Gift!)

11) Write 10^x and $(0.5)^x$ in the form e^{kx}. (Use a calculator.)

12) Find the inverse of $2e^{x-1}$.

13) Solve $2\ln(x + 1) - 1 = 0$.

14) Solve $\ln x - \ln \sqrt{x} - \dfrac{1}{2} = 0$.

15) The common logarithm is used to measure earthquake intensity on the Richter scale. The Richter scale describes the intensity, I, of an earthquake as the number $\log I - \log I_0$ where I_0 is the intensity of a small "benchmark" earthquake. Write the Richter scale rating as a single logarithm. What is the Richter scale rating for an earthquake with intensity $I = 100{,}000 I_0$?

16) Solve $7.2 e^{0.18t} = 2$.

17) Solve $2 \ln x = 3.7$.

18) Solve $5.7 \ln 2x = 3.1$.

19) Solve $1.5 e^{-2.3t} = 7.1$.

20) Solve $\pi e^{-\pi x} = 1$.

Inverse Trigonometric Functions

D.1 Definition of arcsin x, the Inverse Sine Function

We'd like to explore the question of inverses of the sine, tangent, and secant functions. We'll start with $f(x) = \sin x$. Recall the graph:

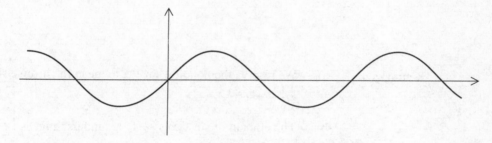

Figure D.1

Right off the bat, we see that $\sin x$ has no inverse, because it fails the horizontal line test (there are horizontal lines that cross the graph more than once). So what can we do? Well, we can consider the **restricted sine function**, defined as $f(x) = \sin x$ for $x \in \left[-\dfrac{\pi}{2}, \dfrac{\pi}{2} \right]$, whose graph is

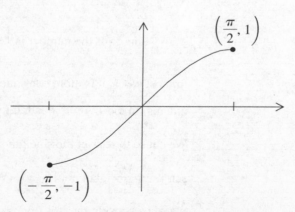

Figure D.2

This function does have an inverse; it is written as **arcsin x** or **sin^{-1} x,** and its graph is obtained by rotating the graph about the line $y = x$.

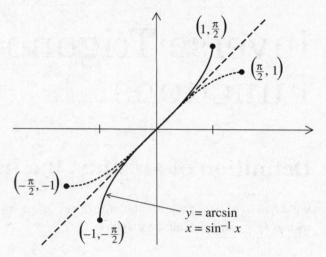

Figure D.3

Remarks

a) We reserve the notation $\sin^{-1} x$ to mean the inverse sine function, NOT
$$\frac{1}{\sin x} = \csc x.$$

b) The domain of $\sin^{-1} x$ is $[-1, 1]$, and its range is $\left[-\dfrac{\pi}{2}, \dfrac{\pi}{2}\right]$, as can be seen from the graph.

c) Recall that $y = \sqrt[3]{x}$ is the inverse of $f(x) = x^3$ and was defined as the number whose cube is x, that is, the number y such that $f(y) = y^3 = x$. In the same way, $y = \sin^{-1} x$ is the number y such that $f(y) = \sin y = x$; but notice that y is restricted to $\left[-\dfrac{\pi}{2}, \dfrac{\pi}{2}\right]$, and so:

$$\boxed{\ \sin^{-1} x \text{ is the number in } \left[-\frac{\pi}{2}, \frac{\pi}{2}\right] \text{ whose sine is } x\ }.$$

In most cases, it is more convenient to think of $\sin^{-1} x$ as the angle (in radians) in the 1st or 4th quadrant whose sine is x.

d) We chose to restrict the sine function to $\left[-\dfrac{\pi}{2}, \dfrac{\pi}{2}\right]$. We could have chosen $\left[\dfrac{\pi}{2}, \dfrac{3\pi}{2}\right]$, but not $[0, \pi]$. (Why not?) But even though the choice is merely a convention, it is important that we all use the same convention.

EXAMPLE 1 Find if possible:

        a) $\sin^{-1}0$

        b) $\sin^{-1}1$

        c) $\arcsin\dfrac{1}{2}$

        d) $\arcsin\dfrac{-\sqrt{3}}{2}$

        e) $\sin^{-1}2$

Solution a) $\sin^{-1}0 = 0$ from the graph.

        b) $\sin^{-1}1 = \dfrac{\pi}{2}$ from the graph.

        c) $\arcsin\dfrac{1}{2} =$ the angle in the 1st or 4th quadrant whose sine is $\dfrac{1}{2}$. Consider the unit circle shown below:

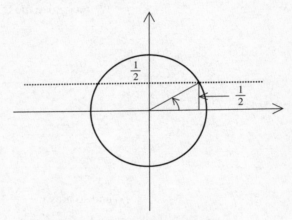

We recognize the 30°/60°/90° triangle:

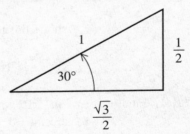

and so $\arcsin\dfrac{1}{2} = \dfrac{\pi}{6}$. A quick glance at the graph of $\sin^{-1}x$ shows that this answer is reasonable.

d) $\arcsin\dfrac{-\sqrt{3}}{2}$ = the angle in the 1ˢᵗ or 4ᵗʰ quadrant whose sine is

$-\dfrac{\sqrt{3}}{2}$. Again, we check the unit circle:

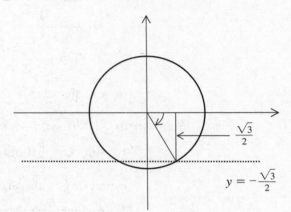

We recognize the triangle,

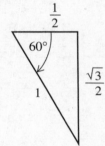

and so $\arcsin\dfrac{-\sqrt{3}}{2} = -\dfrac{\pi}{3}$. Notice we say $\dfrac{-\pi}{3}$, not $\dfrac{5\pi}{3}$, since

$\arcsin\dfrac{-\sqrt{3}}{2}$ is in $\left[-\dfrac{\pi}{2}, \dfrac{\pi}{2}\right]$. Again, a look at the graph of $\arcsin x$

shows this to be reasonable.

e) $\sin^{-1}2$ is not defined, because there is no angle whose sine is 2.   ■

EXAMPLE 2 Find: a) $\sin^{-1}(0.7)$
   b) $\arcsin(-0.913)$
   c) $\sin^{-1}(\pi)$

Solution a) There is no handy triangle that we know of whose side is 0.7, and so we
need to use a calculator. **First make sure that the calculator is set for
radians.** Enter the number 0.7, and then take the inverse sine. On some
calculators this could mean hitting the ⌐2ⁿᵈ⌐ or ⌐INV⌐ button, then the

$\boxed{\sin}$ button. On other calculators there is a $\boxed{\sin^{-1}}$ button. When you do this you should get the approximate answer 0.775, which is consistent with the graph of $\sin^{-1} x$. (Always check.)

b) Using the above procedure, we get $\arcsin(-0.913) \cong -1.151$, which also agrees with the graph.

c) Using the above procedure we get an error message. But this makes sense. Make sure you know why! ∎

$\boxed{\text{D.1}}$ Exercises

1) Evaluate:

a) $\sin^{-1}(-1)$

b) $\sin^{-1}\left(\dfrac{\sqrt{3}}{2}\right)$

2) Evaluate:

a) $\sin^{-1}\left(-\dfrac{1}{2}\right)$

b) $\sin^{-1}\left(-\dfrac{\sqrt{3}}{2}\right)$

3) Find an approximation, rounded to three decimal places, for:

a) $\sin^{-1}(0.8)$

b) $\sin^{-1}(-0.99)$

4) Find an approximation, rounded to three decimal places, for:

a) $\sin^{-1}(0.3)$

b) $\sin^{-1}(-0.01)$

5) Graph:

a) $y = \dfrac{\pi}{2} + \arcsin x$

b) $y = \arcsin(x + 1)$

6) Graph:

a) $y = \dfrac{\pi}{2} - \sin^{-1} x$

b) $y = \sin^{-1}(x - 2)$

$\boxed{\text{D.2}}$ The Functions arctan x and arcsec x

The graph of $\tan x$ also fails the horizontal line test, so we limit ourselves to the restricted tangent function $f(x) = \tan x$ for $x \in \left(-\dfrac{\pi}{2}, \dfrac{\pi}{2}\right)$, with the following graph:

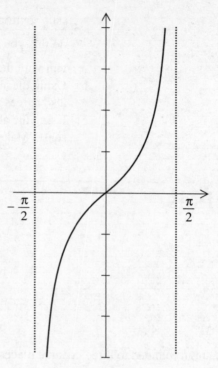

Figure D.4

Hence its inverse $\tan^{-1} x$, or arctan x, looks like this:

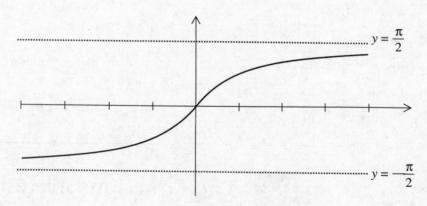

Figure D.5

Notice that:

$$\tan^{-1} x \text{ is the number in } \left(-\frac{\pi}{2}, \frac{\pi}{2} \right) \text{ whose tangent is } x \quad .$$

EXAMPLE 1   Find: a) $\tan^{-1}0$

b) $\tan^{-1}1$

c) $\tan^{-1}\left(-\sqrt{3}\right)$

d) $\tan^{-1}\left(-\sqrt{177}\right)$

Solution a) $\tan^{-1}0 = 0$ from the graph.

b) $\tan^{-1}1$ is the number in $\left(-\dfrac{\pi}{2}, \dfrac{\pi}{2}\right)$ whose tangent is 1, which means that the sine and cosine of that number are equal. This happens only for the angle of 45°. Hence

$$\tan^{-1}1 = \frac{\pi}{4} \cong 0.785.$$

c) $\tan^{-1}\left(-\sqrt{3}\right)$ is a little tougher to recognize. We know that $\tan^{-1}\left(-\sqrt{3}\right)$ is the number in $\left(-\dfrac{\pi}{2}, \dfrac{\pi}{2}\right)$—or the angle in the 1st or 4th quadrant—whose tangent is $-\sqrt{3}$. Checking Example 1(d) of Section D.1, we see that

$$\tan\left(-\frac{\pi}{3}\right) = \frac{\sin\left(-\dfrac{\pi}{3}\right)}{\cos\left(-\dfrac{\pi}{3}\right)} = \frac{-\dfrac{\sqrt{3}}{2}}{\dfrac{1}{2}} = -\sqrt{3},$$

and so

$$\tan^{-1}\left(-\sqrt{3}\right) = -\frac{\pi}{3} \cong 1.047.$$

d) What about $\tan^{-1}\left(-\sqrt{177}\right)$? Well, no triangle is going to help us out, so we use a calculator:

$$\tan^{-1}\left(-\sqrt{177}\right) \cong \tan^{-1}(-13.304) \cong -1.496. \quad \blacksquare$$

The last inverse trigonometric function often used in calculus is the function $\sec^{-1} x$. We consider the restricted secant function $f(x) = \sec x$ for $x \in \left[0, \dfrac{\pi}{2}\right) \cup \left(\dfrac{\pi}{2}, \pi\right]$.

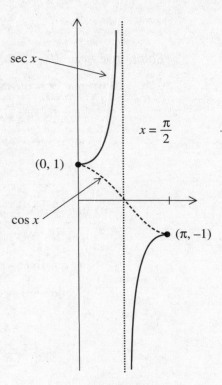

Figure D.6

Its inverse has the graph:

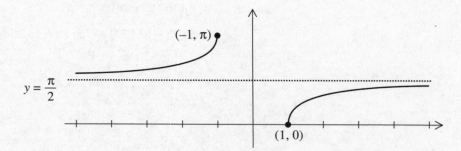

Figure D.7

So $\sec^{-1} x = $ the number in $\left[0, \dfrac{\pi}{2}\right) \cup \left(\dfrac{\pi}{2}, \pi\right]$ whose secant is x. Notice that this strange set is just $[0, \pi]$ without the point $\dfrac{\pi}{2}$.

EXAMPLE 2   Find: a) $\sec^{-1}2$

b) $\sec^{-1}(-1.37)$

Solution a) $\sec^{-1}2 =$ the number in $\left[0,\dfrac{\pi}{2}\right) \cup \left(\dfrac{\pi}{2},\pi\right]$ whose secant is 2, that is, whose cosine is $\frac{1}{2}$. Consider the unit circle shown below and the familiar triangle to the right:

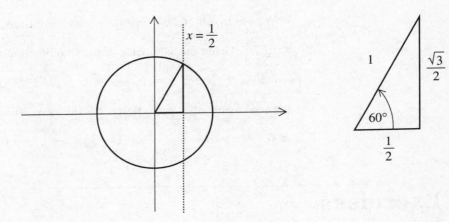

Hence $\sec^{-1}2 = \dfrac{\pi}{3}$.

b) $\sec^{-1}(-1.37) =$ the number in $\left[0, \dfrac{\pi}{2}\right) \cup \left(\dfrac{\pi}{2}, \pi\right]$ whose cosine is $-\dfrac{1}{1.37}$. First of all, $-\dfrac{1}{1.37} \cong -0.7299$. Make sure your calculator is in radian mode. The result will be 2.389. Hence

$$\sec^{-1}(-1.37) \cong 2.389,$$

which agrees with the graph. ■

Remark   The inverses of cos x, cot x, and csc x don't really come up in calculus.

EXAMPLE 3   Solve $3 \sin x = 2$.

Solution We have $\sin x = \dfrac{2}{3}$. The picture looks like:

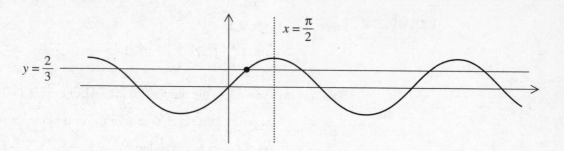

We see that there are infinitely many solutions, one of which is $x = \sin^{-1}\left(\dfrac{2}{3}\right)$ $\cong 0.7297$. Given the symmetry of the sine function about $x = \dfrac{\pi}{2}$, a second solution between $x = \dfrac{\pi}{2}$ and $x = \pi$ is $x = \pi - 0.7297$. The remaining solutions can be obtained using the periodicity of the sine function and can be written as $x \cong 0.7297 + 2k\pi$ and $x = (2k + 1)\pi - 0.7297$ for any integer k. ∎

D.2 Exercises

1) Find the exact value of:

   a) $\tan^{-1}(-1)$

   b) $\sec^{-1}(-2)$

2) Find the exact value of:

   a) $\sec^{-1}\left(\dfrac{2}{\sqrt{3}}\right)$

   b) $\tan^{-1}(\sqrt{3})$

3) Find an approximation rounded to three decimal places for:

   a) $\tan^{-1}(1000)$

   b) $\tan^{-1}(-1000)$

   c) $\sec^{-1}(1.1)$

   d) $\sec^{-1}(-3)$

4) Find an approximation rounded to three decimal places for:

   a) $\sec^{-1}(1000)$

   b) $\sec^{-1}(-1000)$

   c) $\tan^{-1}(0.5)$

   d) $\tan^{-1}(-3.8)$

5) Graph:

   a) $y = \dfrac{\pi}{2} + \tan^{-1} x$

   b) $y = \sec^{-1}(x + 1)$

6) Graph:

   a) $y = \dfrac{\pi}{2} - \sec^{-1} x$

   b) $y = \dfrac{2}{\pi}\tan^{-1} x$

D.3 Inverse Trigonometric Identities

You might expect that $\sin(\sin^{-1} x) = x$ and $\sin^{-1}(\sin x) = x$. Well, not always, because $\sin^{-1} x$ is the inverse of the **restricted sine function**, not $\sin x$. Let's examine this further:

1) Since $\sin^{-1} x$ is the number in $\left(-\dfrac{\pi}{2}, \dfrac{\pi}{2} \right)$ whose sine is x, we see that
$\sin(\sin^{-1} x) = x$ as long as $x \in [-1, 1]$, which is the domain of $\sin^{-1} x$. For other values of x, $\sin(\sin^{-1} x)$ does not exist.

2) Similarly, $\sin^{-1}(\sin x) = x$ for all x in $\left(-\dfrac{\pi}{2}, \dfrac{\pi}{2} \right)$, the domain of the restricted sine function. For other values of x, $\sin^{-1}(\sin x)$ will still exist, but not be equal to x! For example:

$$\sin^{-1}(\sin \pi) = \sin^{-1}(0) = 0 \neq \pi.$$

The situations for the inverse tangent and inverse secant functions are similar:

1) $\tan(\tan^{-1} x) = x$ for all x, and $\tan^{-1}(\tan x) = x$ as long as $x \in \left(-\dfrac{\pi}{2}, \dfrac{\pi}{2} \right)$.

2) $\sec(\sec^{-1} x) = x$ as long as $x \leq -1$ or $x \geq 1$, and $\sec^{-1}(\sec x) = x$ as long as
$x \in \left[0, \dfrac{\pi}{2} \right) \cup \left(\dfrac{\pi}{2}, \pi \right]$.

EXAMPLE 1 Find: a) $\sin\left(\sin^{-1}\dfrac{1}{2} \right)$, $\sin(\sin^{-1} 2)$

b) $\sin^{-1}\left(\sin \dfrac{\pi}{4} \right)$, $\sin^{-1}(\sin 2\pi)$

c) $\tan^{-1}(\tan 2)$

Solution a) The domain of $\sin^{-1} x$ is $[-1, 1]$, and so $\sin\left(\sin^{-1}\dfrac{1}{2} \right) = \dfrac{1}{2}$, while $\sin(\sin^{-1} 2)$ is undefined.

b) The domain of the restricted sine function is $\left(-\dfrac{\pi}{2}, \dfrac{\pi}{2} \right)$, and so
$\sin^{-1}\left(\sin \dfrac{\pi}{4} \right) = \dfrac{\pi}{4}$, while $\sin^{-1}(\sin 2\pi) = \sin^{-1} 0 = 0 \neq 2\pi$.

c) The domain of the restricted tangent function is $\left(-\dfrac{\pi}{2}, \dfrac{\pi}{2} \right)$, so it does not contain 2. Using a calculator, $\tan 2 \cong -2.185$, and so
$\tan^{-1}(\tan 2) \cong \tan^{-1}(-2.185) \cong -1.142.$ ∎

In calculus the inverse trig functions arise mostly in the context of trig substitutions. We sometimes see expressions like $\tan(\sin^{-1} x)$ that can be simplified.

EXAMPLE 2 Simplify $\tan(\sin^{-1} x)$.

Solution Think of $\sin^{-1} x$ as an angle. It is the appropriate angle whose sine is x. Draw a right triangle and label an angle $\theta = \sin^{-1} x$. Then, since $\sin\theta = x$, we recognize that we can get this result by choosing the opposite side to be x and the hypotenuse to be 1. This is shown below:

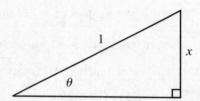

Now, using the Pythagorean Theorem, we see that the base of this triangle is $\sqrt{1 - x^2}$. Hence

$$\tan(\sin^{-1} x) = \tan\theta = \frac{x}{\sqrt{1 - x^2}}.$$

Of course this holds only where both functions are defined, in this case for $x \in (-1, 1)$. ∎

EXAMPLE 3 Simplify $\cos(\tan^{-1} x)$.

Solution In this case, $\theta = \tan^{-1} x$, and the triangle should look like this:

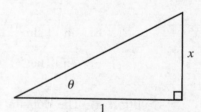

We label the triangle as shown since we wish $\tan\theta = x$. The hypotenuse is $\sqrt{1 + x^2}$, and so $\cos(\tan^{-1} x) = \dfrac{1}{\sqrt{1 + x^2}}$, for all x. ∎

D.3 Exercises

1) Evaluate, with exact solutions if possible; if the expression is undefined, say why:

   a) $\sin(\sin^{-1}(0.1))$

   b) $\tan(\tan^{-1}3)$

   c) $\sec^{-1}\left(\sec\dfrac{2\pi}{3}\right)$

   d) $\sin^{-1}\left(\sin\dfrac{4\pi}{3}\right)$

2) Same as above for:

   a) $\tan(\tan^{-1}(-2))$

   b) $\sin\left(\sin^{-1}\dfrac{4}{\pi}\right)$

   c) $\sin^{-1}\left(\sin\dfrac{4}{\pi}\right)$

   d) $\sec^{-1}\left(\sec\dfrac{1}{2}\right)$

3) Simplify, and indicate where the simplification is valid:

   a) $\cos(\tan^{-1}x)$

   b) $\sec(\sin^{-1}x)$

   c) $\tan(\sin^{-1}x)$

   d) $\cos(2\sin^{-1}x)$
       (*Hint:* Use the fact that $\cos 2x = \cos^2 x - \sin^2 x$.)

4) Same as above for:

   a) $\sin(\tan^{-1}x)$

   b) $\tan(\sec^{-1}x)$

   c) $\sec(\sin^{-1}x)$

   d) $\sin(2\tan^{-1}x)$
       (*Hint:* Use the fact that $\sin 2x = 2\sin x \cos x$.)

The Binomial Theorem

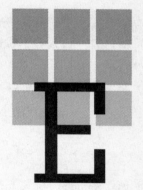

A **binomial** is the sum of two terms and can therefore be represented as $a + b$. The binomial theorem states what happens when powers of $a + b$ are multiplied out. Let's expand $(a + b)^n$ for $n = 1, 2, 3, 4$.

$$(a + b)^1 = a + b$$
$$(a + b)^2 = a^2 + 2ab + b^2$$
$$(a + b)^3 = a^3 + 3a^2b + 3ab^2 + b^3$$
$$(a + b)^4 = a^4 + 4a^3b + 6a^2b^2 + 4ab^3 + b^4$$

We see some obvious patterns. The powers of a decrease as you go from left to right, starting with a^n and ending with a^0 (which equals 1, so it doesn't appear). The powers of b go in reverse, from $b^0 = 1$ to b^n. But what about the coefficients? Let's write them in the form known as Pascal's triangle:

$$
\begin{array}{ccccccc}
 & & & 1 & & 1 & \\
 & & 1 & & 2 & & 1 \\
 & 1 & & 3 & & 3 & & 1 \\
1 & & 4 & & 6 & & 4 & & 1
\end{array}
$$

Each line begins with 1, followed by the line number, n. Also we see that each entry of a particular line is formed by adding the two entries diagonally above it. For example, in the fourth line, 4 is the sum of $1 + 3$, 6 is the sum of $3 + 3$, etc. To get the next line, notice that $1 + 4 = 5, 4 + 6 = 10$, etc., which means that the next line is 1 5 10 10 5 1.

Binomial Theorem

Let n be a positive integer, and let a and b be any numbers. Then

$$(a + b)^n = (\ )a^n + (\ )a^{n-1}b + (\ )a^{n-2}b^2 + \cdots + (\ )a^2b^{n-2} + (\ )ab^{n-1} + (\ )b^n,$$

where the coefficients in the empty parentheses are given by the entries of the nth line of Pascal's triangle.

EXAMPLE 1 Expand $(a + b)^6$.

Solution Check Pascal's triangle. The fifth and sixth lines are

$$1 \quad 5 \quad 10 \quad 10 \quad 5 \quad 1$$
$$1 \quad 6 \quad 15 \quad 20 \quad 15 \quad 6 \quad 1.$$

Hence $(a + b)^6$ is equal to

$$a^6 + 6a^5b + 15a^4b^2 + 20a^3b^3 + 15a^2b^4 + 6ab^5 + b^6. \quad \blacksquare$$

EXAMPLE 2 Expand $(x + 2)^5$.

Solution Check Pascal's triangle. The fifth line is

$$1 \quad 5 \quad 10 \quad 10 \quad 5 \quad 1.$$

Hence $(x + 2)^5$ is equal to

$$(1)x^5 + (5)x^42 + (10)x^32^2 + (10)x^22^3 + (5)x2^4 + (1)2^5,$$

which can be simplified to give

$$(x + 2)^5 = x^5 + 10x^4 + 40x^3 + 80x^2 + 80x + 32. \quad \blacksquare$$

EXAMPLE 3 Expand $(a - 3)^4$.

Solution $(a - 3)^4 = (a + (-3))^4$ (Tricky!)

$$= a^4 + (4)a^3(-3) + (6)a^2(-3)^2 + (4)a(-3)^3 + (1)(-3)^4$$

$$= a^4 - 12a^3 + 54a^2 - 108a + 81. \quad \blacksquare$$

Remarks   a)   This method is clearly simpler than multiplying $a - 3$ by itself repeatedly, but it can become cumbersome if the number n is too big.

b)   The coefficients, which appear in Pascal's triangle, can also be expressed by a formula. This can be found in calculus books in the section on binomial series.

EXAMPLE 4   Simplify $\dfrac{(x + h)^4 - x^4}{h}$.

Solution Check Pascal's triangle. The fourth line is

$$1 \quad 4 \quad 6 \quad 4 \quad 1.$$

Hence

$$(x + h)^4 = (1)x^4 + (4)x^3h + (6)x^2h^2 + (4)xh^3 + (1)h^4$$
$$= x^4 + 4x^3h + 6x^2h^2 + 4xh^3 + h^4$$

and

$$\frac{(x + h)^4 - x^4}{h} = \frac{4x^3h + 6x^2h^2 + 4xh^3 + h^4}{h}.$$
$$= 4x^3 + 6x^2h + 4xh^2 + h^3.   \blacksquare$$

Remark   If we expand $(x + h)^n$, we get $x^n + nx^{n-1}h + (\ )h^2 + (\ )h^3 + \cdots + h^n$, where the open brackets contain polynomials **in x only,** without any factors of h. This fact will be crucial when finding the derivative of x^n.

E.1 Exercises

1) Expand $(x + 1)^6$.

2) Expand $(y - 1)^6$.

3) Expand $(2z + 3)^4$.

4) Expand $(x + \Delta x)^5$.

5) Using the result in Exercise 4, simplify $\dfrac{(x + \Delta x)^5 - x^5}{\Delta x}$.

APPENDIX

Derivation of the Quadratic Formula

In Chapter 2 we demonstrated how to complete the square in a variety of different situations involving quadratic functions. One of its more powerful uses is the derivation of the quadratic formula.

EXAMPLE 1   Complete the square to solve $ax^2 + bx + c = 0$.

Solution First let us consider the algebraic expression $ax^2 + bx + c$. If we wish to complete the square for this general quadratic polynomial we first need to factor out the a from the first and second terms in the expression. That is

$$ax^2 + bx + c = a\left(x^2 + \frac{b}{a}x\right) + c.$$

Now, we need to complete the square inside the parentheses. The coefficient of the x-term is $\frac{b}{a}$, half of that is $\frac{b}{2a}$, and squaring $\frac{b}{2a}$ gives $\frac{b^2}{4a^2}$. So we need to add and subtract $\frac{b^2}{4a^2}$ inside the parentheses. Doing so we get

$$ax^2 + bx + c = a\left(x^2 + \frac{b}{a}x + \frac{b^2}{4a^2} - \frac{b^2}{4a^2}\right) + c.$$

Removing the last term from the parentheses gives:

$$ax^2 + bx + c = a\left(x^2 + \frac{b}{a}x + \frac{b^2}{4a^2}\right) - a\frac{b^2}{4a^2} + c.$$

$$= a\left(x + \frac{b}{2a}\right)^2 - \frac{b^2}{4a} + c$$

$$= a\left(x + \frac{b}{2a}\right)^2 - \frac{b^2 - 4ac}{4a}.$$

If $ax^2 + bx + c = 0$, then from the last equation we have

$$a\left(x + \frac{b}{2a}\right)^2 - \frac{b^2 - 4ac}{4a} = 0.$$

185

This equation is equivalent to:

$$a\left(x + \frac{b}{2a}\right)^2 = \frac{b^2 - 4ac}{4a}.$$

Let's "peel the onion" to solve for x. (See Section 3.1 if you're not sure how to do this.) First get rid of the coefficient a:

$$\left(x + \frac{b}{2a}\right)^2 = \frac{b^2 - 4ac}{4a^2}.$$

Take square roots of both sides:

$$x + \frac{b}{2a} = \pm\sqrt{\frac{b^2 - 4ac}{4a^2}} = \pm\frac{\sqrt{b^2 - 4ac}}{\sqrt{4a^2}}.$$

Now, if $a > 0$, $\sqrt{4a^2} = 2a$, and $\pm\dfrac{\sqrt{b^2 - 4ac}}{\sqrt{4a^2}} = \pm\dfrac{\sqrt{b^2 - 4ac}}{2a}$.

And if $a < 0$, $\sqrt{4a^2} = \sqrt{4}\sqrt{a^2} = 2|a| = 2(-a) = -2a$, and so,

$$\pm\frac{\sqrt{b^2 - 4ac}}{\sqrt{4a^2}} = \pm\frac{\sqrt{b^2 - 4ac}}{-2a} = \pm\frac{\sqrt{b^2 - 4ac}}{2a}.$$

Hence, in both cases we have

$$\pm\frac{\sqrt{b^2 - 4ac}}{\sqrt{4a^2}} = \pm\frac{\sqrt{b^2 - 4ac}}{2a}.$$

So:

$$x + \frac{b}{2a} = \pm\frac{\sqrt{b^2 - 4ac}}{2a}.$$

Subtracting $\dfrac{b}{2a}$ from both sides of this equation gives

$$x = -\frac{b}{2a} \pm \frac{\sqrt{b^2 - 4ac}}{2a}$$

$$= \boxed{\frac{-b \pm \sqrt{b^2 - 4ac}}{2a}}.$$

Look familiar? ■

Answers to Odd-Numbered Exercises

Chapter 1

Exercises 1.1 page 3

1) -19 3) 46 5) $4xy - x + 2y$

7) $3xy - 6x - xy^2 + 2y$ 9) $x^2y + xy^3 - 4x^2y^2 + 2y^3$

11) a) 27 in 8 computations.

b) 27 in 11 computations, so method a) is "cheaper."

Exercises 1.2 page 6

1) $\dfrac{1}{4}$ 3) $\dfrac{3}{5}$ 5) $\dfrac{1}{6}$ 7) $\dfrac{9}{14}$ 9) $\dfrac{21y + 14}{3y}$ 11) $\dfrac{y}{y - 2}$

13) $\dfrac{(x + y)^2}{x}$

15)

```
          (b)   (a)      (c)
    +--+--+--●--+--●--+--●--+--+--+--→ x
   -5 -4 -3 -2 -1  0  1  2  3  4  5
```

Exercises 1.3 page 9

1) $\dfrac{7}{12}$ 3) $\dfrac{7}{30}$ 5) $\dfrac{37}{30}$ 7) $\dfrac{22}{45}$ 9) $\dfrac{29}{42}$ 11) $\dfrac{x + y}{xy}$

13) $\dfrac{4yz - 2xz + xy}{xyz}$ 15) $\dfrac{yz - z(x + 1) + y(x - 2)}{xyz}$

17) $\dfrac{x(z - xy)}{y(x - z)}$ 19) $\dfrac{w - st}{s - 2tw}$ 21) $\dfrac{4y^3z^2 - 2xz^2 + xy}{x^2y^2z}$

Exercises 1.4 page 12

1) $\dfrac{225}{16}$ 3) $\dfrac{60}{19}$ 5) $\dfrac{4}{81}$ 7) $\dfrac{65}{2} = 32\frac{1}{2}$ 9) yz^8 11) x

13) $y^{-12} = \dfrac{1}{y^{12}}$ 15) $\dfrac{x^2}{y}$ 17) $\dfrac{x^6z - x^2}{y^5z^5}$ 19) $\dfrac{xy}{x + y}$

21) $x^{-4} = \dfrac{1}{x^4}$

Exercises 1.5 page 15

1) 12 3) -4 5) -2 7) $\dfrac{2}{7}$ 9) 32 11) 4 13) $\dfrac{27}{64}$

15) 1000 17) $2^{34/15}$ 19) $3^{1/2}$ 21) $3^6 = 729$ 23) $8x^4$

25) $y^{1/15}$

27) No! Let $a = b = 1$; $\sqrt{2}$ is certainly not equal to 2.

29) No! Let $x = 1$; $\sqrt[3]{-7}$ is not equal to -1.

31) $5x^2$ 33) $8x^9$

35) $\dfrac{1}{\sqrt{81x^2 - 4y^2}}$ which is definitely **not** equal to $\dfrac{1}{9x - 2y}$.

Exercises 1.6 page 18

1) 50 3) $1304.444\ldots$ 5) a) \$12.75   b) \$1124.25

7) \$34.00   9) \$191.20   11) \$52,500.00

Exercises 1.7 page 21

1) a) 3.83×10^5 b) -7.24×10^{-4} c) 3.00

d) 2.00×10^2

3) a) 9.48×10^7 b) -3.09×10^{-13}

c) 3.50×10^{-100} d) 7.66×10^{-2}

e) -1.68×10^{-8} f) 5.57×10^{-2}

5) a) $8017.84   b) $3714.87, 48.6% increase.

7) a) 1.83×10^9 b) 6.22×10^{10} c) 34   d) 3299%

Exercises 1.8 page 23

1) a) $[-1, 3]$

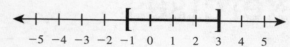

b) $(-1, 3]$

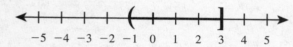

c) $[-3, 1)$

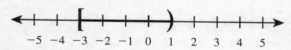

d) $[-3, 4]$

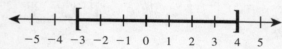

e) $\left(-\frac{1}{2}, \sqrt{2}\,\right]$

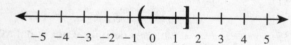

f) $[\pi, 5]$

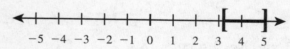

g) $(-\infty, -4)$

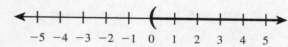

h) $(0, \infty)$

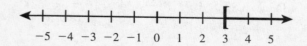

i) $[3, \infty)$

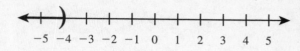

j) $[3 - \pi, \infty)$

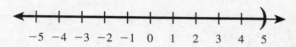

k) $(-\infty, 5)$

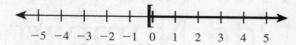

l) $(-\infty, 3]$

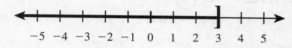

3) a) $3 < x < 7$ b) $-4 < x \le -1$ c) $x \le 19$

d) $2 \le x < 10$ e) $-2 \le x \le -1$

5) a) $[3, 5)$ b) $(-\infty, \infty)$ c) $\varnothing$ empty set.   d) $[4, 7]$

e) $\varnothing$ f) $\{5\}$

Chapter 2

Exercises 2.1 page 28

1) a) $f(x) = (x - 3)^2 + 6$

b) $h(y) = \left(y + \frac{5}{2}\right)^2 - \frac{25}{4}$

c) $g(s) = (s + 1)^2 - 9$

d) $k(x) = 2\left(x - \frac{1}{2}\right)^2 + \frac{9}{2}$

e) $f(x) = 3\left(x - \frac{7}{6}\right)^2 - \frac{37}{12}$

f) $w(x) = \pi\left(x + \frac{1}{\pi}\right)^2 - \frac{1}{\pi}$

3) a) $\left(x - \frac{3}{2}\right)^2 - \frac{77}{4} = 0$

b) $-3(x + 1)^2 + 18 = 0$; alternatively,
$(x + 1)^2 - 6 = 0$

5) a) $\left(x + \dfrac{3}{2}\right)^2 + 2(y - 2)^2 = \dfrac{41}{4}$

b) $3(x + 1)^2 - 2(y + 2)^2 = -16$

c) $-(x - 2)^2 + (y - 8)^2 = 100$

d) $9(x - 2)^2 + 4(y + 1)^2 = 40$

e) $(x - 3)^2 + (y + 5)^2 = 0$ (It's a point!)

7) a) Center $(2, 1)$, radius 4.

b) Center $(3, -2)$, radius π.

c) Center $(-1, -2)$, radius $\sqrt{15}$.

Chapter 3

Exercises 3.1 page 35

1) $\dfrac{4}{3}$ 3) 84 5) $\dfrac{33}{40}$ 7) $\dfrac{75}{7}$ 9) $\dfrac{1}{5}$ 11) $\dfrac{-z^2}{3y^2}$ 13) $-\dfrac{3z + 1}{2z}$

15) $-\dfrac{1}{2y + 3}$ 17) $\dfrac{y^2 + 3y + 1}{2y^2 - 1}$ 19) 15 21) \$600

Exercises 3.2 page 39

1) $-1, -4$ 3) Repeated root, -3. 5) Repeated root, 4.

7) $1 \pm i$ 9) $1, -4$ 11) $-1, 5$ 13) $\dfrac{-7 \pm \sqrt{17}}{4}$

15) $-3 \pm 2\sqrt{3}$ 17) $-1 \pm \sqrt{5}$

19) $-y \pm \sqrt{3}y$, or $-\left(1 + \sqrt{3}\right)y$, and $-\left(1 - \sqrt{3}\right)y$

21) $2(y + z) \pm 2\sqrt{2yz}$ 23) 340 ft

25) a) 40 ft or $26\dfrac{2}{3}$ ft.

b) Height is 32 ft twice, once on the way up, once on the way down.

27) 1 in.

Exercises 3.3 page 45

1) $3, -6$ 3) $-\dfrac{1}{2}, 2$ 5) $3, -3, -6$ 7) $\pm\sqrt{5}, \pm 4$

9) $\pm i, 12$ 11) ± 2, each root is repeated.

13) $\pm 2\sqrt{2}, \pm 2\sqrt{2}i$ 15) $\pm 2, \pm\sqrt{5}$ 17) 16

19) $\dfrac{1}{4}$ or $\dfrac{1}{16}$. 21) 4 23) 1 25) No solution.

27) $-1, 3$ 29) $-\dfrac{5}{2} \pm \dfrac{\sqrt{33}}{2}$

Chapter 4

Exercises 4.1 page 51

1) a) 12 b) 33 c) $(x + h)^3 + 2(x + h)$

d) $8x^3 + 4x$ e) $-x^3 - 2x$

f) $(2 + \Delta x)^3 + 4 + 2\Delta x$

3) a) $|0| = 0, |\pm 1| = 1, |\pm 2| = 2$, etc. so:

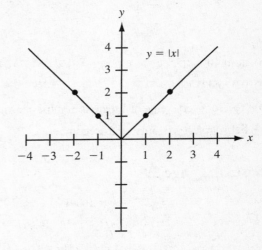

b) $(-\infty, -2) \cup (2, \infty)$

c) domain $= (-\infty, \infty)$, range $= [0, \infty)$.

5) $(-\infty, -3) \cup (3, \infty)$

7) a) $[0, \infty)$

b)

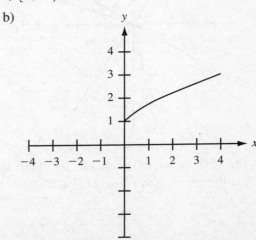

c) $[1, \infty)$

9) a)

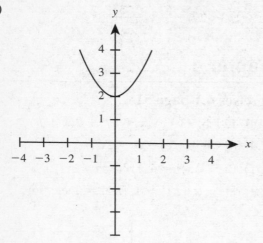

b) domain $= (-\infty, \infty)$, range $= [2, \infty)$.

11) a) 0.3 and 0.1.

b) Yes, for every value of x there is at most one value P.

c) February, April, July, and December.

Exercises 4.2 page 59

1)–5)

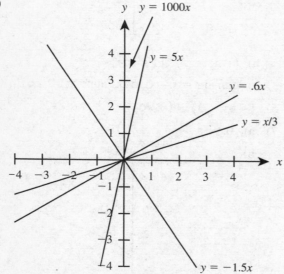

Believe it or not, the graph for $y = 1000x$ is really plotted on this diagram, but because of the

scale you can't really see it. If we scale this as in the next figure you see this but lose the others:

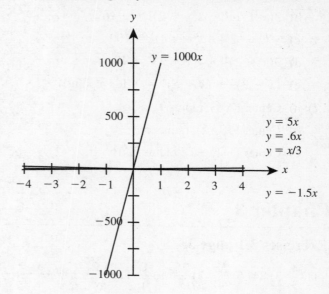

7) $y - 2 = -6(x + 3)$ 9) $y - 4 = -(x + 1)$

11) $y - 4 = 2(x - 1)$, so we get $y = 2x + 2$ and the graph below:

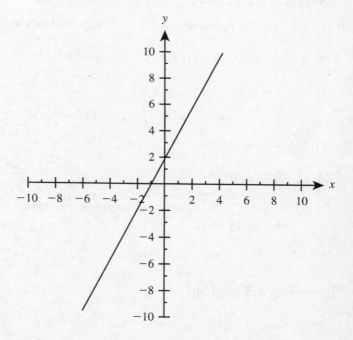

13) $y = -3x + 1$

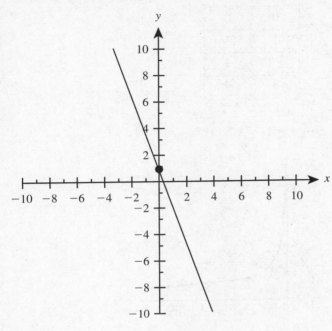

17) $y + 3 = 7\left(x - \dfrac{1}{2}\right)$

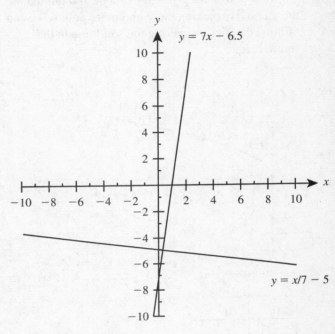

15) $y = 5x + \dfrac{4}{3}$

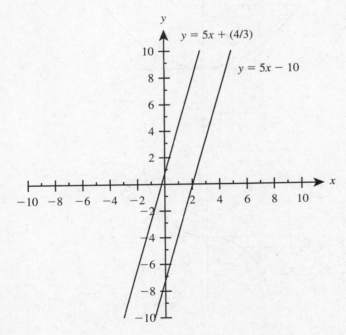

19) a) $F = \dfrac{9}{5}C + 32$

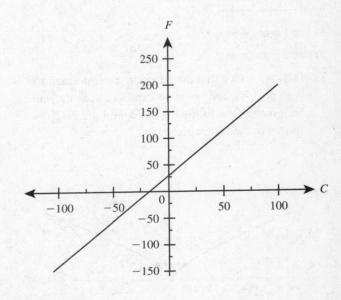

b) 113° F

Exercises 4.3 page 64

1) Our four points are given in the table and plotted on the graph. By picking more and more points, we can "fill in" the picture, getting the solid line in the figure below:

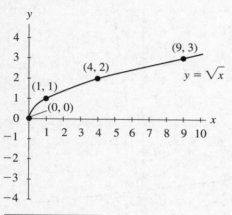

x	$\sqrt{x}$
0	0
1	1
4	2
9	3

3) See Figure 4.8.

5) See Figure 4.11.

7) First we notice that the function $x^{2/3}$ is defined for both positive and negative values of x. Some values are given in the following table, and the graph is shown in the following figure.

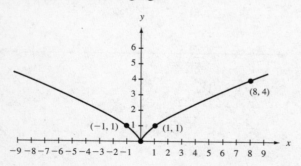

x	$x^{2/3}$
0	0
1	1
−1	1
8	4

9) $y = 1/\sqrt{x}$ 11) $y = ax^n$

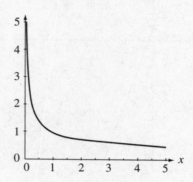

11) $y = ax^n$

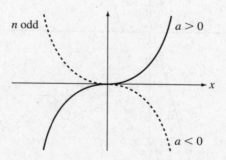

Exercises 4.4 page 66

1)

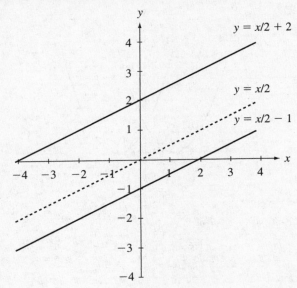

5) a), b)

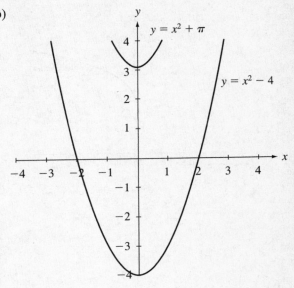

3)

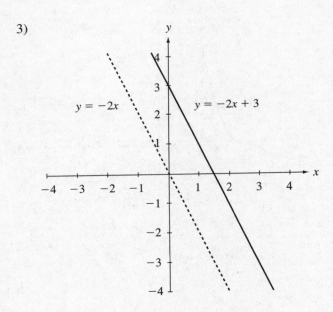

c)

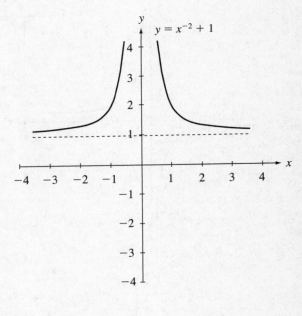

7)

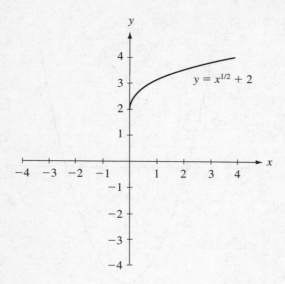

Exercises 4.5 page 69

1)

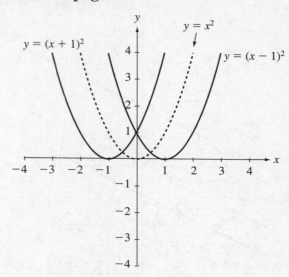

9)

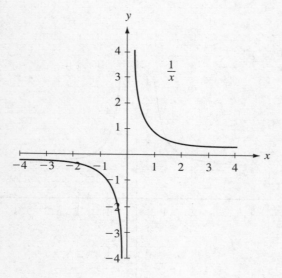

3) a)

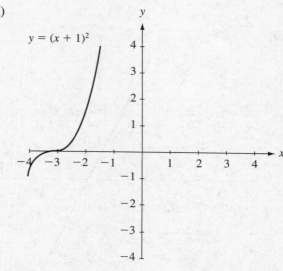

3) b)

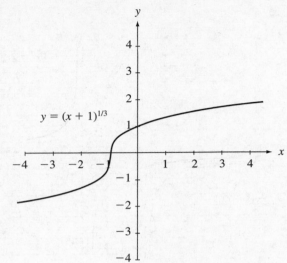

$y = (x + 1)^{1/3}$

5)

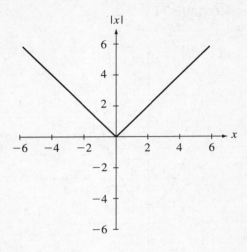

3) c)

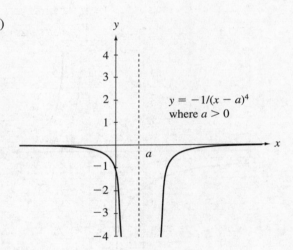

$y = -1/(x - a)^4$
where $a > 0$

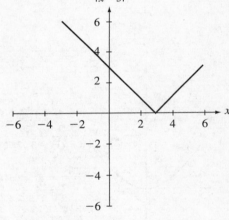

7) $g(t) = [0.6] \cdot [B(t - 18)]$

Exercises 4.6 page 71

1)

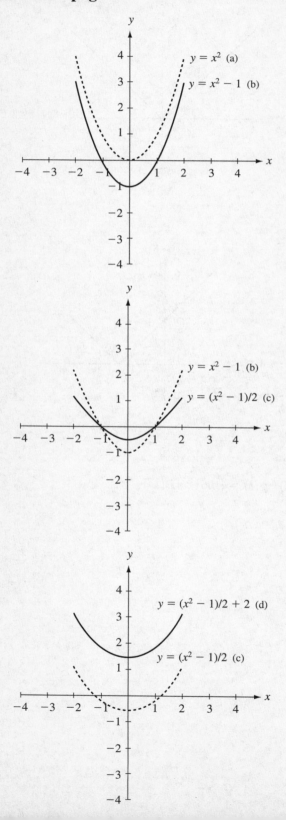

3)

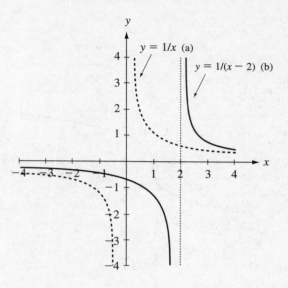

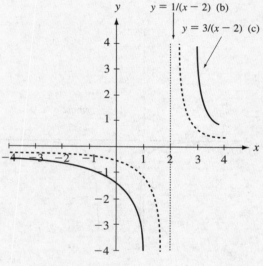

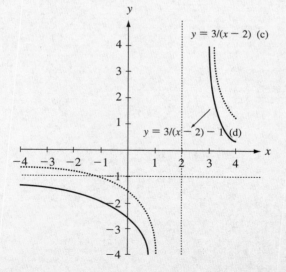

5)

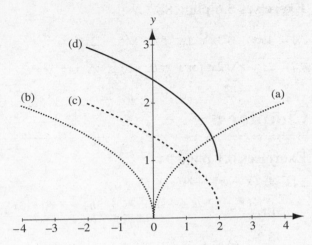

11)

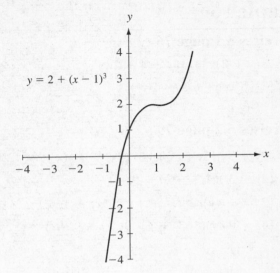

$y = 2 + (x - 1)^3$

7)

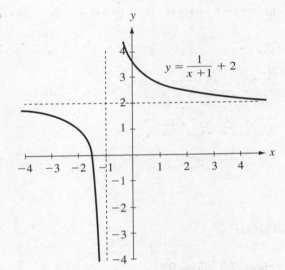

$y = \dfrac{1}{x + 1} + 2$

Exercises 4.7 page 73

1) a) $x = 2$, $y = -5$

   b) $x = 2$, $y = -\dfrac{1}{2}$ and $x = -6$, $y = \dfrac{15}{2}$.

3) $\left(\dfrac{1}{\sqrt{2}}, \dfrac{1}{\sqrt{2}} \right)$ and $\left(\dfrac{-1}{\sqrt{2}}, \dfrac{-1}{\sqrt{2}} \right)$.

5) $\left(\dfrac{1}{2}, \dfrac{\sqrt{3}}{2} \right)$ and $\left(\dfrac{1}{2}, \dfrac{-\sqrt{3}}{2} \right)$.

7) $\left(1 + \sqrt{3}, 2 \right)$ and $\left(1 - \sqrt{3}, 2 \right)$.

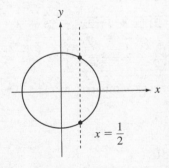

$x = \dfrac{1}{2}$

9)

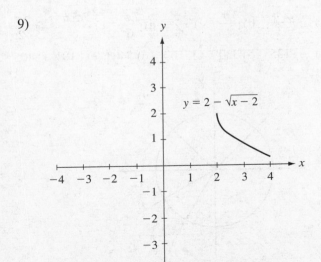

$y = 2 - \sqrt{x - 2}$

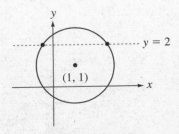

$y = 2$

$(1, 1)$

Chapter 5

Exercises 5.1 page 76

1) $2x(y + 2)$ 3) $x(y + 4 + 2w)$

5) $5x^2y^4(2x^6y^2 + 5 + 4xy^6)$

Exercises 5.2 page 79

1) $(2y + 3z)(2y - 3z)$ 3) $(4x^2 - y^3)(4x^2 + y^3)$

5) $(2s + 3t)(4s^2 - 6st + 9t^2)$

7) $(2^{1/3}x + 4y)(2^{2/3}x^2 - 4 \cdot 2^{1/3}xy + 16y^2)$

9) $(4z - 9^{1/3}t)(16z^2 + 4 \cdot 9^{1/3}zt + 9^{2/3}t^2)$

11) $(x + 1)^2$

13) $(x - 6)(x + 4)$

15) $(s - 2)(s^2 + 2s + 4)(s + 1)(s^2 - s + 1)$

17) $(3x - 2)(x + 1)$ 19) $-2(x - 2)^2$

Exercises 5.3 page 80

1) $(3x + 2y)(a + b)$ 3) $(x - y)(x + 1)(x^2 - x + 1)$

5) $(x^4 + y^2)(x^2 + y)(x^4 - x^2y + y^2)$

7) $(\sqrt{3}x + \sqrt{2}y)(\sqrt{3}x - \sqrt{2}y)(2y)(x + 2)$

9) $(x + y)(3x + 5y + 7)$

Exercises 5.4 page 83

1) $(x - 1)^2(x + 2)$ 3) $(x + 1)(2x^2 - 2x + 3)$

5) $\left(x - \dfrac{3 + \sqrt{17}}{2}\right)\left(x - \dfrac{3 - \sqrt{17}}{2}\right)$

7) Not factorable over real numbers.

9) $(x - 2)(x + 1)^2$

Exercises 5.5 page 86

1) $7(\sqrt{2} + 1)$ 3) $\dfrac{3(x + \sqrt{7})}{x^2 - 7}$

5) $x + \sqrt{3}$ 7) $(x^4 + 3)(x^2 - \sqrt{3})$

Exercises 5.6 page 88

1) $4|x|$ 3) $3x\sqrt[3]{2x}$ 5) $x^2\sqrt{5 + 3x^4}$

7) $2\pi y^2 x\sqrt{2x}$, for $x \geq 0$.   9) $|xy|\sqrt[4]{x + x^2y^6}$

Chapter 6

Exercises 6.1 page 94

1) a) $(x + h)^2 + 3(x + h)$

b) $\dfrac{((x^2 + 2xh + h^2) + 3x + 3h) - (x^2 + 3x)}{h}$

$= 2x + h + 3$

3) a) $2(x + h)^3 - (x + h)$ b) $6x^2 + 6xh + 2h^2 - 1$

5) $\dfrac{1}{x(x - 1)}$ 7) $\dfrac{2x}{(x^2 - 1)}$ 9) 2 11) $\dfrac{x^4 + 3x^2 + 1}{x^3 + 2x}$

13) $\dfrac{a - 9b}{75a^5b^7}$ 15) $\dfrac{1}{\sqrt{x + \Delta x - 3} + \sqrt{x - 3}}$

17) $\dfrac{-1}{\sqrt{x}\sqrt{x + h}(\sqrt{x} + \sqrt{x + h})}$

19) $\dfrac{d(t + \Delta t) - d(t)}{\Delta t}$

Chapter 7

Exercises 7.1 page 98

1) a) $\dfrac{2\pi}{3}$ b) $\dfrac{3\pi}{2}$ c) $\dfrac{3\pi}{4}$ d) $\dfrac{7\pi}{6}$ e) $\dfrac{-5\pi}{6}$ f) $\dfrac{5\pi}{2}$

3) a) $135°$ b) $330°$ c) $-60°$ d) $540°$ e) $810°$ f) $405°$

5)

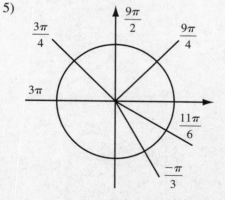

Exercises 7.2 page 104

1) a) -1 b) 0 c) 0 d) 0

3) a) 1 b) 0 c) 0

   d) -1 if k is odd, and 1 if k is even.

5) a) 0 b) -1

Exercises 7.3 page 108

1) $\dfrac{1}{\sqrt{2}}$ 3) $\dfrac{\sqrt{3}}{2}$ 5) $\dfrac{-1}{\sqrt{2}}$ 7) $\dfrac{\sqrt{3}}{2}$ 9) $-\dfrac{\sqrt{3}}{2}$

11) $-\dfrac{\sqrt{3}}{2}$

Exercises 7.4 page 112

1)

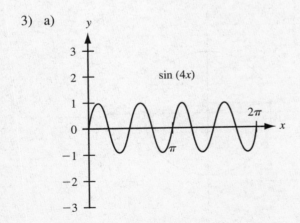

3) a)

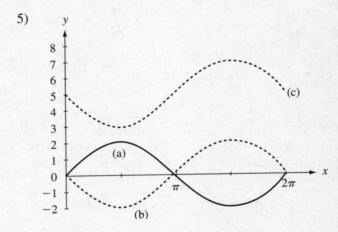

b)

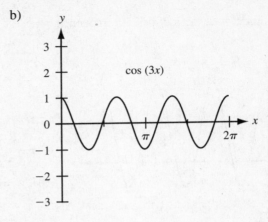

c)

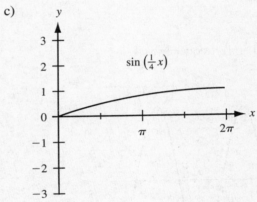

5)

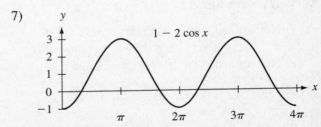

7)

9) Amplitude $= 2$, period $= \dfrac{2\pi}{3}$, frequency $= \dfrac{3}{2\pi}$.

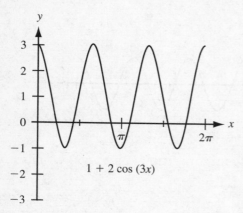

$1 + 2\cos(3x)$

11) Period $= 1$, frequency $= 1$.

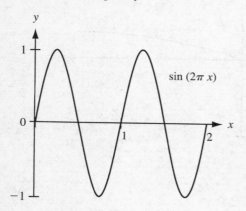

$\sin(2\pi x)$

13)

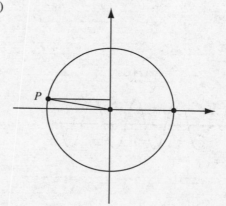

$\sin(-x)$ $\sin x$

5)

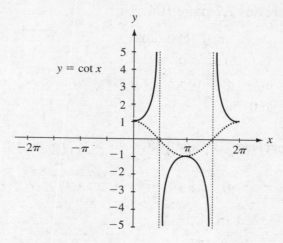

$y = \cot x$

7)

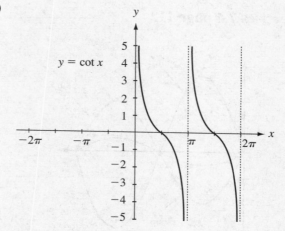

$y = \cot x$

9)

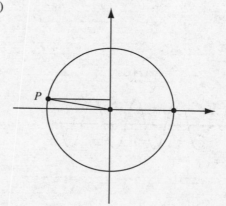

P

3.0 is a very small amount less than π, so $\sin(3.0) = $ 2nd coordinate of $P \approx .1$.

Exercises 7.5 page 114

1) a) Not defined.   b) 1   c) Not defined.   d) 0

 e) Not defined.   f) $\dfrac{\sqrt{3}}{2}$ g) Not defined.   h) 0

3) a) -1 b) 2   c) -2 d) -1

 e) $-\sqrt{3}$ f) $-\sqrt{2}$ g) $-\sqrt{2}$ h) $-\dfrac{1}{\sqrt{3}}$

Chapter 8

Exercises 8.1 page 118

1) a) $\cos^3 x$ b) $(2t + 1)^3$ c) $(\sin x - 4x)^3$
   d) $\cos x^3$ e) $\cos(\cos x)$ f) $\cos(\sin x - 4x)$

3) a) $\tan x - 2\sqrt{\tan x}$ b) $\tan^2 \sqrt{x} - 2\tan\sqrt{x}$
   c) $\tan^{1/2}(x^2 - 2x)$ d) $\sqrt{\tan^2 x - 2\tan x}$
   e) $(x - 2\sqrt{x})^2 - 2(x - 2\sqrt{x})$ f) $\sqrt[4]{x^2 - 2x}$

Exercises 8.2 page 119

1) Outer function $f(x) = x^2$, inner function
   $g(x) = \tan x$.

3) Outer function $f(x) = x^2$, inner function
   $g(x) = x^3 - 1$.

5) Outer function $f(x) = \cos x$, inner function
   $g(x) = x^5$.

7) Outer function $f(x) = \sin x$, inner function
   $g(x) = \sqrt{x}$.

9) Outer function $f(x) = x^{2/3}$, inner function
   $g(x) = \sqrt[5]{x} - 1$.

11) $f(x) = x^3$, $g(x) = \tan 2x$

Chapter 9

Exercises 9.1 page 123

1) -1 3) $\dfrac{-x}{3y}$ 5) $\dfrac{-y(2 + 3xy^2)}{x(1 + 3xy^2)}$ 7) $\dfrac{x(1 + 2y)}{y^2 - 4x^2}$

Chapter 10

Exercises 10.1 page 129

1) Let L = the length of the fencing, and x = the
   width of the field in feet. Then $L = 2x + \dfrac{20{,}000}{x}$ ft.

3) Let P = the perimeter of the window, A = the area
   of the window, and x = the width of the window.
   Then $P = 5x + \dfrac{\pi}{2}x$, and $A = x^2(2 + \pi/8)$.

5) If we let V = the volume of the box and x = the
   length of the square cut out, both in inches, then
   $V = 4x^3 - 36x^2 + 80x$ in.^{3}.

7) $V = \dfrac{S^{2/3}}{6\sqrt{\pi}}$ and $S = \sqrt[3]{36\pi}\, V^{2/3}$.   9) $F \cong 139.8$ lb

11) $A = \dfrac{x^2}{4\pi} + \dfrac{(2 - x)^2}{18}$ 13) 10 ft/hr

15) a) $\dfrac{m}{60}$ b) $2 + \dfrac{4}{3} + \dfrac{m}{300}$ c) 250 mi

Exercises 10.2 page 132

1) $z = \sqrt{34} \cong 5.8$, $\alpha \cong 31°$, and $\beta \cong 59°$.

3) $\beta = 50°$, $y \cong 11.9$, and $z \cong 15.6$.

5) $\beta = 58°$, $z \cong 9.4$, and $x \cong 5$.

7) $\alpha = 36°$, $y \cong 6.9$, and $z \cong 8.5$.

9) The maximum vertical extent of the ladder is approximately 45.1 ft, so the highest it can reach is 49.1 ft. The
   floor level of the sixth floor is 45 ft, so the top of the
   ladder is roughly in the middle of the window. Perfect!

11) Rowing time is $\dfrac{3}{2} \sec \theta$, walking time is
    $\dfrac{1}{4}(10 - 3 \tan \theta)$, so the total time is
    $\dfrac{1}{4}(6 \sec \theta + 10 - 3 \tan \theta)$.

Exercises 10.3 page 135

1) $C = 30°$, $c \cong 7.83$, and $a \cong 15.43$.

3) $A \cong 41.4°$, $B \cong 55.8°$, and $C \cong 82.8°$.

5) The Pythagorean theorem.

Chapter 11

Exercises 11.1 page 140

1) $\cos x(1 - \sin^2 x)^3$ 3) $\pm\sec^2 x(\sqrt{1 + \tan^2 x})^3$

5) $\sec x \tan x(\sec^2 x - 1)^2$ 9) $\tan 2A = \dfrac{2 \tan A}{1 - \tan^2 A}$

11) $\cot 2A = \dfrac{\cot^2 A - 1}{2 \cot A}$

13) $\csc\left(\dfrac{\pi}{2} - \theta\right) = \dfrac{1}{\sin\left(\dfrac{\pi}{2} - \theta\right)}$

$$= \dfrac{1}{\sin\dfrac{\pi}{2}\cos\theta - \cos\dfrac{\pi}{2}\sin\theta}$$

$$= \dfrac{1}{\cos\theta} = \sec\theta.$$

15) a) $\dfrac{1}{4} - \dfrac{1}{2}\cos 2x + \dfrac{1}{4}\cos^2 2x$

b) $\dfrac{3}{8} - \dfrac{1}{2}\cos 2x + \dfrac{1}{8}\cos 4x$

Appendix A

Exercises A.1 page 144

1)

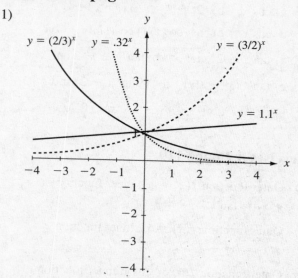

As $x \to \infty$, $.32^x$ and $\left(\frac{2}{3}\right)^x$ both approach 0, while $\left(\frac{3}{2}\right)^x$ and 1.1^x both approach ∞.

As $x \to -\infty$, $.32^x$ and $\left(\frac{2}{3}\right)^x$ both approach ∞, while $\left(\frac{3}{2}\right)^x$ and 1.1^x both approach 0.

3)

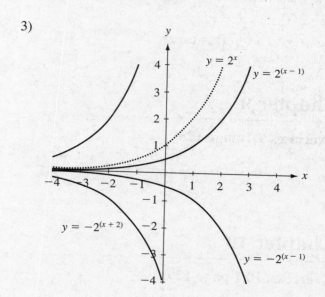

5) a) $\cos(3^x)$ b) $3^{\cos x}$ c) $\cos(3^{\cos x})$

7) a) $f(x) = 2^x, g(x) = x - 1$

b) $f(x) = 3^x, g(x) = \sin x$

c) $f(x) = 5^x, g(x) = \cos(x^2)$

Exercises A.2 page 145

1)

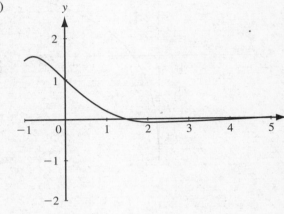

3)

$y = e^{-x^2}$

5) a) $e^{\tan x}$ b) $\tan(e^x)$ c) $\cos(e^{e^x})$

7) a) e^{3x} b) e^x c) $e^x + 1$

9) a) $f(x) = e^x, g(x) = 3x + 1$

b) $f(x) = \cos x, g(x) = e^x$

c) $f(x) = \sin x, g(x) = e^{x^2}$

d) $f(x) = \dfrac{1}{x}, g(x) = \cos(e^x)$

e) $f(x) = x^2, g(x) = \cos(e^x)$

11)

13) a) 8.585   b) .0000454 $= 4.54 \times 10^{-5}$ c) 1.609

Appendix B

Exercises B.2 page 152

1)

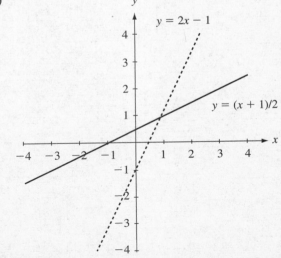

3) The function $g(x)$ is its own inverse.

5)

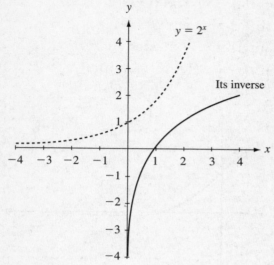

7)

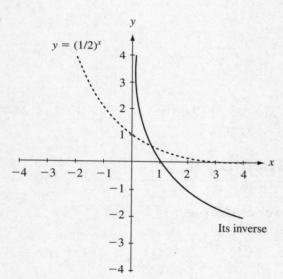

9)

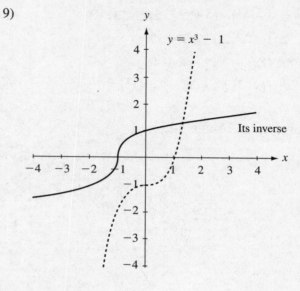

11) a)

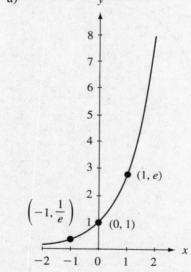

b) It passes the horizontal line test.

c)

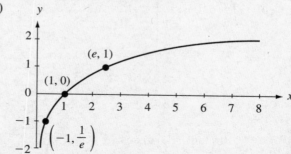

Exercises B.3 page 155

1) $f^{-1}(x) = \dfrac{x+3}{2}$ 3) $g^{-1}(x) = \dfrac{x^3 - 1}{5}$

5) $f^{-1}(x) = \dfrac{2}{x}$

7) It flunks the horizontal line test; no inverse. However, if $x \geq 1$, it passes the test.

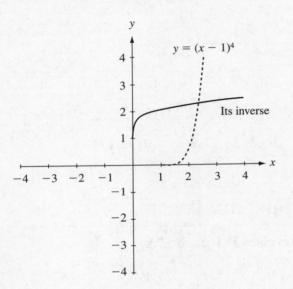

9) a) 15,840   b) 7040   c) 2   d) 5

Appendix C

Exercises C.1 page 157

1) 2   3) −1   5) −3   7) −3   9) $\dfrac{1}{2}$ 11) $\dfrac{3}{4}$ 13) x

15) $\dfrac{1}{100}$ 17) 80   19) 15   21) x

Exercises C.2 page 162

1)

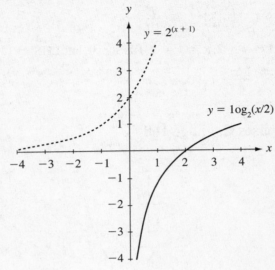

3) a) 9   b) 2   c) $\dfrac{5}{2}$ d) −8

5) $\dfrac{-20}{3}$ 7) ±2   9) $\pm\sqrt{14}$

11)

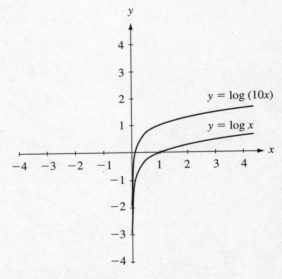

Notice that $\log 10x = 1 + \log x$.

Exercises C.3 page 165

1) 9　3) $\left(\dfrac{1}{10}\right)^{1/3}$　5) $\dfrac{7}{10}$

7) Let $a = 2, x = y = 1$. RHS = 0, but LHS = 1.

9) $\dfrac{\pi}{4}, \dfrac{3\pi}{4}$　11) $0, \pi, 2\pi$

Exercises C.4 page 168

1)

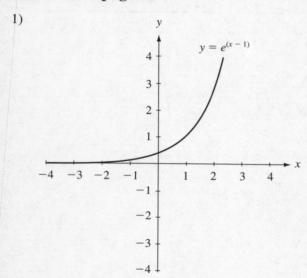

3)

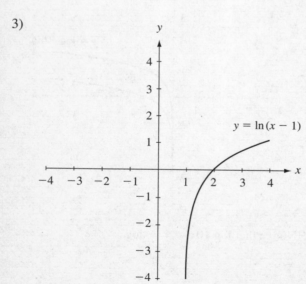

5)

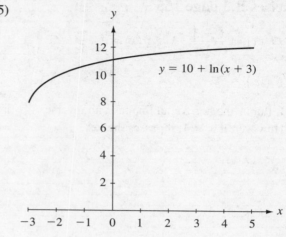

7) e^2　9) $1, -3$　11) $e^{2.3x}, e^{-.69x}$　13) $\sqrt{e} - 1$

15) $\log\dfrac{I}{I_0}, 5$　17) 6.36　19) -0.676

Appendix D

Exercises D.1 page 173

1)　a) $-\dfrac{\pi}{2}$　b) $\dfrac{\pi}{3}$　3) a) 0.927　b) -1.429

5)　a) $y = \dfrac{\pi}{2} + \sin^{-1} x$

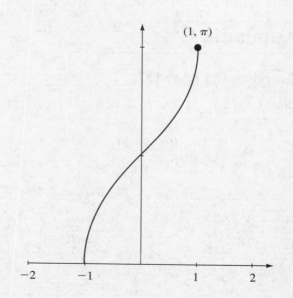

b) $y = \sin^{-1}(x + 1)$

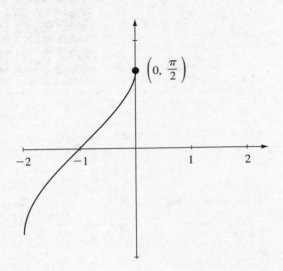

$\left(0, \dfrac{\pi}{2}\right)$

Exercises D.2 page 178

1) a) $-\dfrac{\pi}{4}$ b) $\dfrac{2\pi}{3}$

3) a) 1.570   b) -1.570 c) 0.430   d) 1.911

5) a) $y = \dfrac{\pi}{2} + \tan^{-1} x$

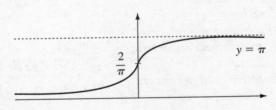

$\dfrac{2}{\pi}$ $y = \pi$

b) $y = \sec^{-1}(x + 1)$

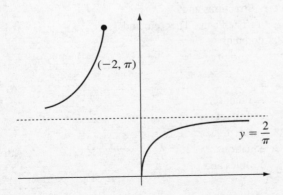

$(-2, \pi)$

$y = \dfrac{2}{\pi}$

Exercises D.3 page 181

1) a) 0.1   b) 3   c) $\dfrac{2\pi}{3}$ d) $-\dfrac{\pi}{3}$

3) a) $\dfrac{1}{\sqrt{1 + x^2}}$, for all x.

b) $\dfrac{1}{\sqrt{1 + x^2}}$, for x in $(-1, 1)$.

c) $\dfrac{x}{\sqrt{1 - x^2}}$, for x in $(-1, 1)$.

d) $1 - 2x^2$, for x in $(-1, 1)$.

Appendix E

Exercises E.1 page 184

1) $x^6 + 6x^5 + 15x^4 + 20x^3 + 15x^2 + 6x + 1$

3) $16z^4 + 96z^3 + 216z^2 + 216z + 81$

5) $5x^4 + 10x^3(\Delta x) + 10x^2(\Delta x)^2 + 5x(\Delta x)^3 + (\Delta x)^4$

INDEX